P9-ELP-088

One Bird Strike and You're Out

Solutions to Prevent Bird Strikes

Dr. Jerry LeMieux

Order this book online at www.trafford.com
or email orders@trafford.com

Most Trafford titles are also available at major online book retailers.

Note for Librarians: A cataloguing record for this book is available from Library and Archives Canada at www.collectionscanada.ca/amicus/index-e.html

Printed in Victoria, BC, Canada.

ISBN: 978-1-4269-2085-1 (sc)

Library of Congress Control Number: 2009939806

Our mission is to efficiently provide the world's finest, most comprehensive book publishing service, enabling every author to experience success. To find out how to publish your book, your way, and have it available worldwide, visit us online at www.trafford.com

Trafford rev. 11/11/2009

North America & international
toll-free: 1 888 232 4444 (USA & Canada)
phone: 250 383 6864 • fax: 812 355 4082

What if the weather had been bad on the day that Flight 1549 launched? It would have been a catastrophe. The government would have reacted much more aggressively in the aftermath, as they did back in '80s when airliner-to-airliner mid-air collisions had become so frequent that there arose a public outcry

According to the Federal Aviation Administration (FAA) database, in the last 19 years there have been 33 accidents where two or more engines of an airliner took bird strikes.....that's almost two per year. Several of these accidents have caused crashes and completely destroyed the aircraft.

On a global scale, aircraft bird strikes occur almost every single day. Many strikes cause damage but no loss of life. It may seem strange to say, but we have been lucky so far; the recent ditching of US Airways Flight 1549 in the Hudson is a warning sign that should not be ignored. Although the pilot made an extremely prudent decision and carried it out with expert skill, the successful outcome of this accident involved a great deal of chance. All 155 passengers and crew survived this accident, but if the weather had been poor or if the clouds had been lower to the ground, the result would have been catastrophic. During the recent NTSB hearings on Flight 1549, testimonies revealed that bird strike collisions with aircraft are a worsening threat.

Expert airline pilot commentary is provided on the complete NTSB transcript of Flight 1549. In this book, Dr. LeMieux shares his own personal experience with a bird strike in a supersonic military jet and delineates survival procedures for airline passengers in the event of a water landing. Every airline passenger should read this book.

This book is based on the firm belief that together, the FAA, US Department of Agriculture (USDA), DoD, pilots, controllers, airport managers and biologists all form an important team. Only through collaboration can we properly analyze the increasingly urgent problem of aircraft bird strikes and discover an efficient and effective solution.

Acknowledgements

I would like to thank my family for their support.

Preface

Equipped with a PhD in electrical engineering, over 22 years of R&D experience, and 35 years (more than 10,000 hours) of flying time in both fighter and commercial aircraft, I've recently filed a patent for an aircraft bird strike warning system. This idea has taken form gradually over the years, but it suddenly became clear to me when US Airways Flight 1549 ditched in the Hudson River upon being hit by a flock geese. As a major airline pilot, I put myself in the shoes of Capt. Sully Sullenburger and began to analyze the course of events that led him to his decision to ditch in the Hudson. Every year, airline pilots go through extensive training so that we will be ready to execute these maneuvers in the event of an emergency. As you'll learn in the chapters ahead, when an aircraft collides with a flock of large birds, a successful outcome only becomes possible when pilot skill combines with a certain element of luck.

In order to increase the likelihood of safety and survival in an aircraft bird strike, a wide range of precautions are being taken. Agencies throughout the nation and all over the globe are currently collaborating to diminish the chances of these accidents occurring; however, the advances being made are few and far between, and it is time for an altogether new and more comprehensive solution.

My extensive experience qualifies me to make insightful and achievable suggestions in this regard. I have enjoyed a very successful second career as an electrical engineer specializing in radar system design, development, integration and testing. On the staff and faculty at MIT, Boston University and Daniel Webster College, a top-ranked aviation school, I have taught courses in electrical and aeronautical engineering as well as advanced mathematics. Furthermore, I have been a systems engineer and program manager responsible for cost, scheduling and technical performance. So, after reading about the FAA's current project to evaluate commercial bird radars for deployment to our nation's airports, I was able to notice that something very critical was missing – namely, the operational considerations associated with warning the pilot of a potential collision with a flock of birds in flight. Some companies call their bird radars "bird strike prevention radars", but in truth, a radar alone cannot prevent a collision between an aircraft and a bird – only a pilot can.

According to hearings on Flight 1549 conducted by the FAA and National Transportation Safety Board (NTSB), the practice of using a radar as a sensor to detect and track birds holds great promise. Currently, though, commercial bird radar companies are taking the approach of displaying bird targets to an air traffic controller so that he can then warn pilots of an impending collision. The first problem with this approach is that busy controllers have no time to monitor yet another display. At some of the busiest US airports in Class B airspace,[1] the number of takeoffs and landings exceeds 1 every 15 seconds; so a bird radar must not create more work for the air traffic controller. The second problem with these systems is that, at these airports, the control tower communication frequency is also extremely busy – in fact, it's often difficult for pilots to get a word in edgewise. With this level of frequency saturation, even if the controller had the time to observe a bird radar display, he would often have no way to reach the pilot with this information.

So part of the equation is missing – namely, how to get the bird target data to the pilot so that he can decide about changing his flight path to avoid a collision. For these reasons, I strongly believe that bird targets need to be sent directly to the cockpit; and fortunately, today, we have the technological capability to do so, although the necessary components have not been assembled into the ideal combination – yet.

---

1 'Class B' means an airport serving over 5 million passengers per year.

INTRODUCTION

The first fatality caused by the collision between an aircraft and a bird occurred in 1912 on a transcontinental flight when a gull jammed in the aircraft control cables of a Wright Flyer, causing loss of control and a crash. Since the time of that accident, many more aircraft have taken to the skies. And since 1912, bird populations have increased dramatically too. In fact, over the past 40 years, bird populations have skyrocketed due to wildlife preservation efforts. It is estimated that since 1990, the goose population alone in the United States has almost quadrupled from 1 to 4 million. And just to give you a big picture here, there are some 650 species of birds in the United States.

For the most part, the largest threat is posed not by single birds but by flocks of large birds.[2] According to the FAA database for aircraft bird strikes, the total number of strikes between 1990 and 2008 was 89,727; however, bird strike reporting is currently not mandatory. In fact, it is estimated that as little as 20% of all bird strikes are reported in commercial aviation. So, according to these figures, there must have been about 448,635 actual bird strikes over this 19-year period. That's an average of 65 bird strikes a day, just within the Untied States!

2 One of the worst air disasters of all time, however, was actually caused by large flocks of small starlings.

It is important to note that bird strike accidents are not just limited to commercial aviation; significant and costly aircraft bird strike accidents have occurred in the military as well. In 2007, over 5,000 bird strikes were reported by the US Air Force (USAF). Bird strikes have caused the complete loss of an E-3 AWACS and a B1B aircraft. Commercial aircraft are limited to 250 knots when below 10000 feet and 200 knots when in controlled airspace around an airport. Although some fighter aircraft require a greater airspeed to allow for safe maneuvering (i.e., speeds of up to 350 mph), most commercial and military jet aircraft are limited to 250 knots below 10000 feet. One of the reasons for this limitation is to reduce the damage caused by the potential impact of an aircraft bird strike.

Over the past 25 years, aircraft collisions with birds have caused many large airliners to crash and hundreds of lives to be lost. It is estimated that aircraft bird strikes annually cost the aviation industry $650 million in the United States and over $1.2 billion worldwide.[3] More than 75% of all aircraft bird strikes occur near airports – most (92%) occur below 3000 feet, during takeoff or landing; three-quarters (72%) below 500 feet; and more than half (63%) below 100 feet.

In the next 10 years, annual air passenger traffic is expected to increase from the current value of 600 million to 1 billion, and over the next 25 years, the projections for increased air travel stand at 40%. On top of all this, jet engines are also being made quieter, which increases the chance that birds will fail to hear these airliners as they approach.

Aircraft bird strikes occur almost every single day. Many strikes cause damage but no loss of life. It may seem strange to say, due to the enormity of the losses that bird strikes have caused, but we have been lucky so far; the recent ditching of US Airways Flight 1549 in the Hudson is a warning sign that should not be ignored. Although the pilot made an extremely prudent decision and carried it out with expert skill, the successful outcome of this accident involved a great deal of chance as well. All 155 passengers and crew

3 For a list of significant worldwide aircraft bird strike accidents in 2009 as well as a history of all significant aircraft bird strikes in the United States, see Appendix 1.

survived this accident, but if the weather had been poor, or if the clouds had been lower to the ground, the result would have surely been catastrophic. During the recent NTSB hearings on Flight 1549, testimonials revealed that bird strike collisions with aircraft constitute a worsening threat.

Clearly, something needs to be done. This book discusses the efforts of those who are currently working on this problem, the approaches they are taking in order to solve it, and the latest technologies available that could be brought together to create a bird strike warning system for pilots.

It is my aim to educate the public on the seriously increasing threat that aircraft bird strikes pose for airline travel. I have gathered information and insights from various fields and agencies together to provide readers with a more comprehensive understanding of the challenges that must be overcome in order to make this bird strike problem a thing of the past.

Contents

Acknowledgements	vii
Preface	ix
Introduction	xi
Chapter 1: Airliner in the Hudson	1
Chapter 2: Ditching an Airliner	19
The Airplane	26
The Crew	27
Chapter 3: FAA Aircraft Bird Strike Database	37
Chapter 4: Aircraft Accident Investigations	55
Chapter 5: Bird-Proofing Airliners	72
The Airframe	75
The Empennage	75
The Windshields and Windows	76
The Structure of the Rest of the Aircraft	76
The Engine	79
Chapter 6: Organizations Working on the Bird Strike Problem	87
NTSB and FAA	87
Air Line Pilots Association (ALPA)	90
Bird Strike Committee USA	92
The International Bird Strike Committee	98
German Bird Strike Committee	98
The International Center for the Study of Bird Migration in Latrun, Israel	99
The UK Bird Strike Committee (UKBSC)	100
The USAF Bird/Wildlife Aircraft Strike Hazard (BASH) Team	101
FAA and USDA	102
FAA, USAF, US Army, US EPA, US Fish and Wildlife Service & the USDA	103
Chapter 7: Controlling Birds at Airports	109
Federal Agencies and Laws that Protect Birds	109
Controlling the Bird Problem at Airports	112

Aircraft Flight Schedule Modification 113
Habitat Modification and Exclusion 114
Repellant Techniques 118
Harrassment Techniques 121
Wildlife Removal 127

Chapter 8: Recent Advances in Bird Radar Technology 132

Chapter 9: BAM, BASH, AHAS and Advanced Concepts 148

Chapter 10: Control Tower Implementation for Bird Radars 170

Chapter 11: Teamwork in Aviation 186
Flight Attendant Communication with Pilots 202
Emergency Procedures 204
Normal Procedures 205
Practices and Procedures 206

Chapter 12: Pilot Workload and the Requirement for Automated Bird Warnings 209

Chapter 13: Cockpit Data Link Implementations for Bird Warnings 225
Digital Automatic Terminal Information System (D-ATIS) 228
VHF Digital Link Mode 2 (VDL-2) 230
NextGen 231
Other Advanced Concepts Internet Using Wi-Fi in the Sky 236
In-Flight Pico Cell Phone Technology 239

Chapter 14: Aircraft Collision Avoidance Systems for Bird Warnings 243

Chapter 15: Recommendations 260
Allocate Long-Term Funding to Develop a Collision Avoidance System for Birds 261
Develop CONOPS for Bird Radars to Warn Pilots & Controllers 263
Approve and Fund Airport Purchases of Bird Radars 266
Improve Bird Radars 268
Train Pilots on the Hazards of Bird Strikes 269

Summary 270
Appendix 1: Significant Worldwide Aircraft Bird Strikes for 2009 & History of Serious Aircraft Bird Strike Accidents in the USA 271
Appendix 2: Turbine Engine Foreign Object Ingestion and Rotor Blade Containment Type Certification Procedures 290
Substantiation Criteria 297
Appendix 3: Bird Ingestion Certification Standards for Jet Engines 299
Background 299
Summary of the NPRM 300
Summary of the Final Rule 300
Summary of Comments 300
The Amendment 302
PART 33—AIRWORTHINESS STANDARDS: AIRCRAFT ENGINES 302
Sec. 33.76 Bird ingestion. 303
Appendix 4: NTSB Aircraft Bird Strike Recommendations from Accident Investigations 307
Appendix 5: AIM Regulations for Ditching an Airliner & Ditching Demonstrations for Aircraft Certification 312
Appendix 6: Commercial Airport Certification (FAR Part 139) 331
FAR Part 139 Subpart B CertificationSec. 139.101 - Certification requirements: General. 331
FAR Part 139 Subpart D OperationsSec. 139.305 - Paved areas. 332
Sec. 139.307 - Unpaved areas. 333
Sec. 139.309 - Safety areas. 334
Sec. 139.311 - Marking and lighting. 335
Sec. 139.313 - Snow and ice control. 337
Sec. 139.319 - Aircraft rescue and firefighting: Operational requirements. 338
Sec. 139.321 - Handling and storing of hazardous substances and materials. 343

Sec. 139.323 - Traffic and wind direction indicators. 345
Sec. 139.325 - Airport emergency plan. 346
Sec. 139.329 - Ground vehicles. 349
Sec. 139.331 - Obstructions. 350
Sec. 139.333 - Protection of navaids. 350
Sec. 139.335 - Public protection. 351
Sec. 139.337 - Wildlife hazard management. 351
Sec. 139.339 - Airport condition reporting. 354
Sec. 139.341 - Identifying, marking, and reporting construction and other unserviceable areas. 355
Appendix 7: TCAS Installation and Use 357
Ground training must cover the following areas: 364

Chapter 1:

Airliner in the Hudson

To obtain the time, altitude, airspeed and conversations that took place in the cockpit, as well as those that took place between the controllers and the pilots, I have used the NTSB transcript and simulation. My professional expertise as a major airline pilot and my experience as an Executive Safety Chairman for the Air Line Pilots Association (ALPA) qualify me to explain the rationale behind the choices made during this crisis, and I have added some commentary to parts of the transcript in order to share this reasoning with you.

On 15 January, 2009, at 3:25 PM, US Airways Flight 1549 takes off from LaGuardia airport in New York. The aircraft makes a left turn and begins climbing to 5000 feet. The pilots are cleaning up the aircraft (i.e., bringing the gear and flaps up), accelerating to 250 knots, and executing the after-takeoff checklist. Two minutes later, at 3:27:09 PM, 219 knots, 2700 ft, the aircraft collides with a large flock of Canada geese.

> **3:27:15, 209 knots, 2856 feet** – A pilot says "Uh oh, we got one roll—both of 'em rolling back." A loss of thrust has just been felt by the pilots, and as the aircraft continues to climb, a loss of speed can be observed on the airspeed indicator. The fuel flow is decreasing and the Exhaust Gas Temperature

(EGT) is increasing toward the redline, indicating insufficient airflow to the engine. Neither engine seems to have flamed out (i.e., quit), but a low RPM is indicated. Insufficient thrust is available to maintain level flight. The passengers heard loud thumps on both sides of the cabin followed by quiet. Some passengers saw flames shooting out of the engines. A few seconds later, there was a little smoke and the foul smell of a bird cooking.

3:27:19, 202 knots, 2932 feet – A pilot says "Ignition start." All pilots are trained to immediately attempt a restart of the engines at this point, so the engine igniters are placed in the restart position and the pilots prepare to attempt a restart of both engines.

3:27:21, 198 knots, 2973 feet – A pilot says "I'm starting the APU." The Auxiliary Power Unit (APU) is a small internal jet engine in the back of the aircraft that can provide an independent source of electrical power. The pilot is starting the APU as a backup, in case the engines fail to restart.

3:27:24, 194 knots, 3021 feet – The Captain says, "My aircraft," and the co-pilot says "Your aircraft." Up this point, the co-pilot had been flying, since this leg of the journey was his; however, when the Captain decides to take control of the aircraft, this verbal communication between the pilots is necessary to verify who is responsible for flying.

3:27:27 PM, 185 knots, 3070 feet – The aircraft continues to climb to 3070 feet and the airspeed slows to 185 knots. The Captain says "Get the QRH…loss of thrust on both engines." When an emergency occurs in flight, pilots are trained to have one pilot fly the aircraft while the other pulls out the Quick Reference Handbook (QRH). This checklist includes a series of steps that are taken to fix the aircraft, at least temporarily, and prepare it for a safe landing. Here, the pilots are trying to control the aircraft, communicate with controllers, and assess the damage. Physical damage to the aircraft is being assessed by looking at the windscreen, windshield wipers and radome. Engine damage is assessed by looking at the engine instruments. A low engine rpm on the engines indicates a loss of thrust.

LaGuardia Controller: "Cactus fifteen forty-nine, turn left heading two seven zero."

3:27:35 PM, 185 knots, 2957 feet – Pilot: "Ah this is, uh, cactus fifteen thirty-nine hit by birds, we lost thrust in both engines, we're turning back towards LaGuardia."

3:27:42 PM, 190 knots, 2725 feet – LaGuardia Controller: "Ok, yeah, you need to return to LaGuardia, turn left heading two two zero."

3:27:46 PM, 192 knots, 2564 feet – The controller was trying to give the pilots directions to land on runway 13 at LaGuardia airport. After rolling out of the turn on this heading, the aircraft is unable to maintain altitude and starts to descend. The pilot then comes over the intercom and says, "This is the Captain, brace for impact." The flight attendants then go through their brace-for-impact speech.

Pilot: "Two two zero."

3:27:47 PM, 195 knots, 2504 feet – LaGuardia Controller: "Tower, stop your departures, we got an emergency returning."

3:27:52 PM, 202 knots, 2276 feet – The co-pilot says "If fuel remaining, mode selector, ignition, ignition." The co-pilot is carrying out the steps in the checklist and verbalizing each step so the other pilot can listen as he flies the aircraft.

LaGuardia Tower: "Who is it?"

LaGuardia Controller: "It's fifteen twenty-nine, he—ah, bird strike, he lost all engines. He lost the thrust in the engines. He is returning immediately."

3:27:58 PM, 208 knots, 1998 feet – The co-pilot says, "Thrust levers confirm idle." Any time the thrust needs to be pulled back during an emergency, both pilots must confirm which engine will be pulled back.

LaGuardia Tower: "Cactus fifteen twenty-nine, which engines."

LaGuardia Controller: "He lost thrust in both engines, he said."

LaGuardia Tower: "Got it."

3:28:05 PM, 214 knots, 1779 feet – Co-pilot: "Airspeed optimum re-light, 300 knots, we don't have that." The co-pilot is verbalizing that the aircraft does not have sufficient airspeed to restart the engines.

LaGuardia Controller: "Cactus fifteen twenty-nine. If we can get it to you, do you want to try to land runway one three?"

3:28:10 PM, 213 knots, 1632 feet – Pilot: "We're unable. We may end up in the Hudson."

3:28:13 PM, 212 knots, 1612 feet – The aircraft is descending at a very slow and controlled rate. The pilots are looking for a place to put the aircraft down. The passengers are calm and very quiet.

LaGuardia Controller: "Alright. Cactus fifteen forty-nine, it's going to be left traffic to runway three one."

Pilot: "Unable."

3:28:35 PM, 200 knots, 1395 feet – Captain: "The left one is coming back up a little." Both pilots are looking at the rpm gauges, hoping for an increase in the indication. The controller is unaware of the severity of the thrust loss and continues to try to vector the pilot to return to LaGuardia airport. Rather than have a discussion with the controller in the midst of this crisis, the pilot summarizes everything into the one-word response "Unable".

LaGuardia Controller: "Okay, what do you need to land . . . Cactus fifteen forty-nine. Runway four is available if you want to make left traffic to runway four."

Pilot: "I am not sure if we can make any runway. Oh, what's over to our right? Anything in New Jersey? Maybe Teterboro."

3:28:54 PM, 198 knots, 1190 feet – LaGuardia Controller: "Okay, yeah. Off to your right side is Teterboro airport. Do you want to try and go to Teterboro?"

Pilot: "Yes." At this point, the pilot probably cannot see the runway but wants to keep all possible options open.

3:29:04 PM, 186 knots, 1167 feet – The co-pilot continues to read the checklist "No re-light after 30 seconds. Engine master one and two confirm off." The pilots are turning the ignition switches to the off position in order to attempt another restart. The pilot is able to see Teterboro airport and has tentatively made a decision to go there. The Captain makes a PA announcement and says "This is the Captain, brace for impact." The flight attendants then said "Brace, brace, heads down, stay down."

LaGuardia Controller: "Teterboro, uh, Empire, actually, LaGuardia departure got an emergency inbound."

Teterboro Controller: "Okay, go ahead."

LaGuardia Controller: "Cactus fifteen twenty-nine over the George Washington Bridge wants to go to the airport right now."

Teterboro Controller: "He wants to go to our airport. Check. Does he need any assistance?"

LaGuardia Controller: "Ah, yes. He, ah—he was a bird strike. Can I get him in for runway one?"

Teterboro Controller: "Runway one. That's good."

LaGuardia Controller: "Cactus fifteen twenty-nine, turn right two-eight-zero. You can land runway one at Teterboro."

Pilot: "We can't do it."

3:29:25 PM, 192 knots, 726 feet – By this time, the pilot must have seen how far away the runway actually was. Using his expert judgment, he decides that he does not have enough altitude and airspeed to glide into Teterboro airport.

LaGuardia Controller: "Okay, which runway would you like at Teterboro?"

Pilot: "We're gonna be in the Hudson."

3:29:29 PM, 192 knots, 657 feet – Here, he makes the final decision to ditch in the Hudson River.

LaGuardia Controller: "I'm sorry. Say again, cactus."

3:29:35, 190 knots, 534 feet – The controller can't believe what he is hearing, so he asks the pilot to repeat what he said. The controller must have also realized that he might be the last person to talk to anyone on the crew.

LaGuardia Controller: "Cactus, ah, cactus fifteen forty-nine, radar contact is lost. You also got Newark airport off your two o'clock and about seven miles."

3:29:46, 189 knots, 312 feet – Here, the controller is trying to maintain communication; he just lost contact with the aircraft as it dropped under his radar coverage. The co-pilot says "No relight," and the Captain says, "Ok, let's go. Put the flaps out." The flaps generate more lift for the aircraft at a slow airspeed.

LaGuardia Controller: "Cactus fifteen twenty-nine, uh, you still on?"

3:30:16 PM, 139 knots, 360 feet – Passengers are placing their heads in their laps.

LaGuardia Controller: "Cactus fifteen forty-nine, you got runway two nine at Newark at your two o'clock and seven miles.

3:30:30 PM, 131 knots, 225 feet – The Captain says "Got any ideas?" and the co-pilot responds "Actually not." Aircrews are trained to work as a team, and the Captain is asking the co-pilot if there is anything they may have missed.

3:30:43 PM, 128 knots, 0 feet – After a hard jolt, the aircraft comes to a stop next to the Intrepid Aircraft Sea, Air and Space Museum. A total of 3 minutes and 34 seconds had elapsed between the time of the bird strike and landing in the Hudson. After the landing, the nose of the aircraft tilts upward due to damage in the rear section of the aircraft that allowed water to enter the cargo compartment. Later, the flight attendant in the forward cabin will describe the touchdown as firm, and the aft flight attendant will describe it as violent. Initially, the flight attendants do not realize they were in the water. No one in the cabin has been hurt, and the passengers remain calm. The fight attendants yell "evacuate" and passengers move toward the doors and get on the rafts. A total of 155 passengers and crewmembers evacuate the aircraft. The evacuation proceeds very smoothly, and it is not long until the passengers are picked up by boats. Many passengers are standing on the wing. By 4:20 PM, all passengers and crewmembers are off the airplane. The aircraft is then tied to tugboats, pushed to Battery Park and tied to a pier.

When a loss of thrust occurs to both engines, airline pilots are trained to react immediately. Pilots are routinely tested, both orally and in the simulator, to make sure the correct response occurs. When emergencies occur during critical phases of flight, or in cases where severe consequences are likely to arise if certain actions are not immediately accomplished, then the first few steps in the QRH checklist must be executed from memory. Once these steps have been executed, the pilot pulls out the checklist and attempts to restart the engines.

During an airborne emergency, the lines of responsibility are normally divided between pilots – one pilot flies the aircraft and talks on the radio, while the other uses the checklist to place the aircraft in a safe configuration to land. This division of responsibility is necessary to ensure that someone is always flying the aircraft. Although this division of responsibilities is necessary, it also leaves room for miscommunication. This may sound like common sense, but many years ago, aircraft accidents occurred because both pilots were trying to fix the problem and no one was flying the aircraft. I can remember one accident in particular where the consequences were fatal. An L-1011 was making an approach into Miami International Airport and the pilots noticed that the nose landing gear had not come down. From the cockpit, pilots can't see the landing gear, so they have to rely on indicators. When the landing gear is extended and down, each gear shows up as a green light on the cockpit display. Well, in this instance, the nose gear light was not green, so the aircraft went around, the autopilot was engaged, and both pilots worked the checklist to try lowering the gear. About halfway around the traffic pattern, the pilots didn't notice that the autopilot altitude hold function had disengaged, and the aircraft began a very gentle descent. The pilots were so consumed in fixing the problem that they failed to realize that no one was flying the aircraft. Eventually, the aircraft impacted the ground in the Florida Everglades, and over 100 people perished. During the investigation, it was discovered that the gear light had not illuminated due to a burned-out light bulb.

Since that tragic accident, the FAA has mandated some major changes. First, all commercial aircraft are required to have a warning system installed that provides a visible light, as well as an audible warning, when the autopilot disengages. Second, pilots are trained to ensure that, when an in-flight emergency occurs, one person is flying the aircraft at all times. The pilot flying still may have input during the execution of the checklist, but his primary responsibility is to fly the aircraft. Pilots are tested in the simulator on a regular basis to ensure that this division of responsibility is executed properly. A video reenactment of this particular scenario was made in the simulator and is typically shown to commercial pilots during training.

Less than 25 years ago, poor crew communication was a contributing factor in aviation accidents. Issues such as poor communication, poor task prioritization, poor decision making and lack of leadership were cited as causes in most commercial aviation accidents. In response to many accident investigations, the NTSB recommended that the FAA require commercial aircraft carriers to make Crew Resource Management (CRM) training mandatory for their aircrews. CRM is a management technique that originated from a NASA workshop; it focuses on reducing pilot error and is incorporated into initial and recurrent pilot training. CRM training is not just limited to aircrews, though; it is taught to flight attendants, maintenance personnel and air traffic controllers as well. The most basic premise of CRM training is to work together as a team when handling abnormal aircraft events. The results of this training are evident in the transcript above – the Captain is flying the aircraft, the co-pilot is helping the Captain, the flight attendants are working with the Captain to prepare the passengers for an impact, and the controllers are trying to help the Captain find a place to land. They all collaborated to handle a very serious aircraft emergency.

Perhaps you're wondering "Why couldn't the pilots get the engines restarted?" Well, I believe that the damage to the engines was so significant that a restart was impossible. FAA regulations require that jet engines be designed to withstand a bird strike – each engine is required to withstand a certain quantity of birds of a certain size. According to the FAA, for the US Airways Flight 1549 aircraft and single bird ingestion, the jet engine must be able to ingest a 4-pound bird and shut down safely. For a flock of birds, the same jet engine must be designed to ingest seven 1.5-pound birds, not lose more than ¼ power, and continue to run for at least 5 minutes.[4]

4 For a complete description of FAA Certification standards for jet engine ingestion of birds, please see Appendix 2.

Canada geese have an average weight of 8 pounds and a maximum weight of 16 pounds. So, in this particular case, the engine ingested birds with a weight exceeding the design limits.

In some accounts that I've read, the engines were still running at a reduced rate (and thus providing electrical power). So, it is possible that the engines survived the bird ingestion but that there was not enough airspeed for the re-light. When jet engines flame out, the manufacturer has a restart envelope that must be followed for a successful restart. This envelope includes an altitude and airspeed window. Every altitude requires a different airspeed for a successful restart. The speeds are normally listed in checklists and onboard flight manuals, and modern aircraft show the correct speed on a cockpit display.

So why did the aircraft end up in the Hudson River when there were several New York airports nearby? The answer to this question is simply 'pilot judgment'. The more hours a pilot has, the better his judgment is. In the United States, through a process that involves constant change, the FAA and the Department of Transportation (DOT) have made flying a very safe mode of transportation. It is when an abnormal event or aircraft emergency occurs that this pilot judgment comes into play. After losing both engines to a bird strike, in addition to flying the aircraft and communicating with air traffic control (ATC), the pilots were continuously evaluating their landing options. They also had significant help from the controllers, who were continuously providing the pilot with landing options. Candidates for landing were to return to LaGuardia or land at Teterboro. In the ATC environment, however, all input, including comments from the controller, are merely suggestions for the pilot to consider. In the end, it is the pilot in command who is responsible and who, therefore, has the final decision authority.

As you can imagine, the pilots were very busy. In the case of dual engine failure, pilots look for an area that is flat so as to prevent collisions with any obstructions on the ground. At high altitudes, though, it is impossible to see detailed terrain features. The decision to land in the Hudson was not made until all other options had been exhausted. The Captain did consider landing at Teterboro, but then it seems that he must have looked inside at the aircraft airspeed and altitude parameters and, based on those measurements, decided that Teterboro was unreachable. If the pilot had decided to go to Teterboro, I believe that he wouldn't have made it. That decision could have been catastrophic.

The amazing outcome – that this accident resulted in no loss of life – can be attributed to the skill of the well-trained crew and a certain element of luck. If the weather had been bad, it's very likely that no one would have survived. When both engines are lost, a secondary source of electrical power, the ram air turbine, automatically deploys and generates a backup source of electrical power. So, if the weather had been bad, the pilots would still have had their instruments, but they would have been unable to see the ground. In the weather, there is limited visibility; you know you are coming down – you just don't know where. The amount of time you have to look for a place to land is directly proportional to how low the ceiling (i.e., the height of the clouds) is. If the ceiling were down to 200 feet, for example, there would be virtually no time to decide on where to land, and the pilot would have had to just accept where he was. The result would most likely have been an impact into a densely populated area with total loss of passengers and crew and probable loss of life on the ground as well.

So, did the pilots do the right thing? With my 35 years and 10,000 hours of both military and commercial flying, it is my expert opinion that the pilots did the exact right thing. The cockpit recordings illustrate that the pilots were executing the correct procedures and working together as a crew to solve the problem. They had excellent control of the aircraft as they descended at the aircraft's maximum glide speed, thereby allowing the aircraft to glide the maximum distance. If you are too fast, you won't go as far. If you are too slow, you could stall the aircraft. They listened to the controller's suggestions, made decisions, and communicated their intentions; they communicated inside the aircraft with the flight attendants and passengers; they never gave up on the possibility of an engine restart. After continuously evaluating their options for where to land, they ultimately arrived at the correct decision to land in the Hudson.

In February, Congress held hearings on the accident. Captain

Sully Sullenberger testified that he had 42 years and 20,000 hours of flying experience and had flown USAF F-4 fighter jets in the 1970s (which is incidentally, the same jet I flew in the 80s). With gratitude, he stated that the successful results of this accident were attributable to the team that included the co-pilot, the flight attendants, the controllers and the quick responders of New York City.

The NTSB testified that the left engine had been knocked off during the water landing and recovered three days later. The engines were then transported for analysis to the GE plant in Cincinnati, Ohio. The rest of the aircraft is in the storage yard in Kearney, NY, where it is to remain for the duration of the investigation. The flight recorders were recovered intact. The NTSB also mentioned that they had been publishing a number of recommendations to the FAA and other agencies regarding bird strikes ever since 1973, when the problem was first identified (see Appendix 4).

Several months later, in June 2009, the NTSB held hearings on the Flight 1549 accident. For those who are new to this process, these hearings are NTSB's way of adding to the facts under consideration in the accident investigation. Also, the NTSB uses the hearings to determine the cause of the accident and share the investigation process with the public. In general, the main objective of the NTSB is to determine the cause of an accident and make recommendations that will prevent that sort of accident from happening again. The hearings focused mostly on pilot training and certification standards for ditching an airliner, bird strike detection and mitigation efforts, and certification standards for jet engine bird ingestion. Testimonials were given by the representatives of US Airways, the aircrew, NASA, the USDA, aircraft manufacturers, the FAA, and the jet engine manufacturers.

US Airways testified on the academic and simulator training that its pilots receive on the handling of a dual engine failure. NASA testified on their Emergency and Abnormal Situations (EAS) study, which aims to improve aviation safety and the safety of space operations. It strives to achieve this by combining our understanding of human learning and performance with our knowledge of emergency, abnormal, and off-nominal situations; it also addresses the operational contexts in which they occur, with the intent of informing the design, evaluation, and implementation of procedures and the training of crews to respond in the most effective ways possible to these situations. This study focused on policies, procedures, practices, checklists, aircraft systems, training, human performance, crew coordination and response as well as others (e.g., instructors, operations personnel, etc.).

The USDA testified that bird strikes annually cost the US aviation industry $650 million ($1.2 billion worldwide) and that, since 1988, 229 lives have been lost. In fact, in 1991, a single Canada goose strike cost an airline $3.5 million. This testimony highlighted the fact that, in the United States, there are 14 bird species with a body mass greater than 8 pounds, and the population of almost every one of these species is increasing. Also mentioned was the fact that several solutions have been proposed, including the improvement of wildlife management programs at airports and the continual evaluation of radars.

The aircraft manufacturers testified on certification requirements for loss of all engines and the structural requirements for the ditching of an airliner.

The FAA testified on the regulations for certification of jet engines in regard to bird ingestion and the wildlife management program that is currently assessing bird radars.

The jet engine manufacturers testified on their compliance and their practice of testing to meet the FAA regulations for bird ingestion. Here, the investigators were trying to establish if the requirements for aircraft and engine design were adequate. If they are deemed inadequate, then the NTSB recommends changes.

For many years, this accident will be studied and scrutinized by NASA, the airlines, the media and academia. The NTSB will make many recommendations for the FAA to create regulations aimed towards improving air safety. Since the cargo area should not have cracked open and filled with water, aircraft design standards are likely to become more stringent to improve survivability in the event of a water landing. Jet engine design standards may also be changed so that they may withstand the ingestion of larger birds. Consequently, bird radars will most likely receive increased recognition and publicity as a potential way to improve air safety. The FAA will continue to evaluate the radars and likely approve them for purchase by airports.

So what goes through a pilot's mind after an aircraft bird strike? It depends on what phase of flight you're in. When I was a T-38 instructor pilot in the USAF, I successfully survived an aircraft bird strike due to my intensive training, quick reaction time and a bit of good luck. The T-38 is a supersonic trainer used to train pilots that are headed to a fighter assignment. USAF pilot training normally takes about a year, and only those students in the top portion of the class are selected to fly this particular aircraft. Flying is a skill that must be constantly practiced in order to be maintained, and one of the maneuvers that is always practiced is takeoffs and landings. In order to perform multiple takeoffs and landings on a single flight, though, the pilot must accomplish a maneuver called a touch-and-go. This occurs when a pilot lands and, after landing, pushes the power up to perform another takeoff. After taking off, the pilot then flies around the airport to practice another landing.

So, on this particular day, the student was flying. I was in the back seat as an instructor pilot, and we were in the traffic pattern. As we approached the runway, the student pulled his power back, and we settled down momentarily on the runway. Soon after, he pushed up the power for another takeoff, and just then, at about 140 knots, we ingested a bird in the right engine. It wasn't obvious that we had experienced a bird strike because, at that speed, you can't see a bird approaching. I simply heard a buzzing sound. When something is not right in an aircraft, you check your engine instruments first because the engines are a very critical component. So when I looked at the engines, I expected to see both at full power because the throttles were advanced full forward. This was not what I saw though – the power setting for right engine was very low.

At this very critical phase of flight, the next thinking process that goes through a pilot's head is whether to continue the takeoff or abort and try to stop the aircraft on the runway. This is one of the most difficult decisions a pilot can make. In an airliner, an incorrect decision can get a lot of people killed. It is crucial moments like these that necessitate annual training in a full-motion simulator.

Normally, the decision is based on a critical engine failure speed, which depends on the weight of the aircraft on that particular flight and is calculated before each takeoff. On their airspeed indicator, pilots usually place a marker on this critical engine failure speed so that, in the event of an engine failure during takeoff, the decision will be simple — if you're below the set airspeed, you abort; if you're above it, you takeoff. This is sometimes called a "go/no-go decision". The reason you takeoff if above the set airspeed is simply because the other option would be too dangerous – that is, if you abort at such a high speed, there won't be enough runway for you to stop the aircraft, and you will roll off the end of the runway.

So if it's such a simple decision, why would pilots ever make incorrect decisions and cause major accidents? Well, first of all, this decision normally has to be made within the space of a few seconds. At this phase of flight, pilots are trying to fly the airplane by looking outside to make sure they are going straight down the runway. They're not staring inside the aircraft at their airspeed indicator. There are about a thousand things a pilot must think about during a takeoff. And in a high-speed jet aircraft, you have to think faster than the aircraft. You must plan ahead and think about the heading you will fly, the configuration changes (gear and flaps) that you will make, and the control inputs you will make in order to fly the departure procedure. You have to listen to the air traffic controllers' instructions; monitor your airspeed, altitude and heading; and make adjustments to stay on course as you maneuver in three dimensions. So you are quite task-saturated during this phase of flight, and while you are trying to fly the aircraft, you may not recognize the problem in time to make the right decision.

In the airlines, luckily, there are two pilots, so you have a better chance of recognizing problems in time to make the right decision. During takeoff, the airspeed indicator is set at the critical engine failure speed, and the pilot not flying actually calls out the airspeed indications. When the set speed is reached, the pilot not flying calls it out so that the one who is flying doesn't have to look inside the aircraft and can, instead, concentrate on flying. If he hears the callout of the decision speed, the pilot flying normally makes the decision to takeoff because he knows there will not be enough runway to stop the aircraft if he aborts after the call.

An engine failure on takeoff is one of the most difficult emergencies a pilot can deal with, even though it may sound simple for a quick decision maker. There are unique circumstances that will cause a pilot to modify the go/no-go decision. One of these anomalies is an engine fire. If you are past the set speed and you have a fire, it may be better to abort and take your chances on the ground. This is because flying around with an aircraft on fire has unpredictable consequences. In this eventuality, pilots are actually trained to takeoff, use onboard engine fire extinguishing systems to put the fire out, shut down the engine, and return for an immediate landing. However, there could be extenuating circumstances that dictate an abort.

Another situation that could cause a pilot to abort above the set speed is an aircraft that is unsafe to fly. An incredible example of this occurred on a TWA flight out of JFK on July 30, 1992. The airliner actually became airborne when the stall warning systems activated. The stall warning system activates when the aircraft can no longer fly as it is approaching a stall, and a crash will likely occur if this happens close to the ground. If a stall warning occurs, a pilot is taught to make an immediate maneuver to correct the situation. In this case, the pilot made the unusual decision to set the aircraft back down on the runway and attempt an abort. Of course, the airliner rolled off the end of the runway and caught fire because it was past the go/no-go speed. Unfortunately, the cause of the stall warning was later determined to be a malfunction of the aircraft stall warning system. A pilot is normally trained to trust his instruments. Of all the abort decisions that have been made throughout history, this one stands out because the aircraft was so far past the go/no-go speed that, even though the pilots knew the aircraft would be unable to stop, they decided to abort based on safety.

There are other events that will cause a pilot to modify his go/no-go decision, which brings us full circle to the conclusion of my personal bird strike story. We were at a very fast airspeed and not actually taking off from the beginning of the runway. During a touch-and-go, the plane is further down the runway from the approach threshold, so when an engine failure occurs, there is less runway available for stopping. It's not like you're doing a normal takeoff. A touch-and-go is a modified takeoff, and there are really no rules for making decisions. Decisions, in this case, are made based on experience. The hours of experience translate directly into pilot judgment. The decision I made was to take the aircraft away from the student and abort. To make this decision within a few seconds, I looked inside at the airspeed indicator and then outside at the remaining markers on the runway to determine how much runway I had left. There are numbered markers along the runway (4 means 4000, 3 means 3000, etc.). Using my pilot judgment, I decided that I had enough runway to stop the aircraft.

Well, I may have been right, but just barely. I used the brakes to slow down, and I stopped on the last brick of the runway. At high speeds, brakes heat up quickly, and if the speed is too high, the tires can actually catch fire and explode. Whether or not this occurs depends on the speed at which the aircraft first applies the brakes. I had a successful result because of my training, quick reaction time, light braking and, of course, a bit of luck. I could have easily ended up in the overrun or off the end of the runway with an aircraft accident on my resumé. Fortunately, though, in over 10,000 flying hours, I have never had a major aircraft accident.

Also worth noting is the method by which a pilot handles an airborne bird strike when the engines are operating. After a bird strike, the first thing a pilot tries to do is assess the damage to the aircraft. This assessment can be accomplished in a number of ways. In the cockpit of an airliner, you can't see very much, so you rely on your instruments. Another great source of information is to ask the flight attendants if they heard or saw anything abnormal from the cabin. In the case of a bird penetrating the windshield, a pilot needs to avoid the distraction of the bird's blood and feathers. If the bird has penetrated the windshield or if the windshield is cracked, it is important to remember to slow the aircraft down to reduce the wind and noise levels. In this case, the pilot should put on sunglasses or smoke goggles to reduce wind effects.

Pilots are trained extensively to cope with these emergencies in preparation for when they might occur, but since so much luck is involved in a successful outcome, our efforts should focus, instead, on prevention. In summary, two main factors are currently contributing to the increasing risk of bird strikes: first, wildlife conservation regulations have caused large increases in bird populations in the United States; and second, airline traffic in this country is expected to increase. Specifically, annual worldwide passenger volume is expected to increase at a rate of approximately 5% per year, from 5 billion today to 11 billion in 2027. Together, these two facts mean that there will be more planes and more birds fighting for the same amount of airspace. Therefore, something needs to be done now to prevent another accident like the ditching in the Hudson.

Chapter 2:

Ditching an Airliner

"The FAA does not specifically require that pilots demonstrate ditching in a simulator."
FAA Fact Sheet published June 9, 2009

I've had a personal experience with nearly bailing out and ditching my fighter in the Atlantic. As a former military fighter pilot, I've been trained to refuel in flight. I was stationed in Europe during the cold war, and there were training requirements for maintaining proficiency. We'd often arrange for a refueling tanker aircraft to deploy to a given location and hookup for gas on the way to the practice areas. So, every once in a while, we'd deploy to the United States to participate in war game exercises at Nellis AFB, Nevada (Red Flag, Green Flag, etc.). To get there, we would normally join up with two tankers off the coast of England – one refueler would take us halfway across the Atlantic then turn around and return to England, while the other tanker would take us the rest of the way across. Normally, there were six fighter aircraft assigned to each tanker.

As we crossed the Atlantic, we would always plan for the worst-case scenario just in case one fighter aircraft was unable to take gas. There were divert bases along the route of flight that we would continuously monitor. Normally, we would watch our fuel in flight, and when it got to a predetermined level, we would take turns on the tanker getting gas. Once all aircraft had received fuel, we could go back to cruising in a formation flight with the tanker.

On one particular deployment, we had planned for seven refuelings. I had the Wing Commander and the Deputy Commander for Operations, who were both Colonels, in my flight. We were halfway across the Atlantic, and it was my turn to refuel. I opened the refueling door and flew into position below the tanker. The tanker flew the refueling probe into position, and I made my connection. As I proceeded to fly formation underneath the tanker, I watched my refueling gauge. Normally, the gauge would increase in quantity. Well, this was not happening, and I was starting to get concerned. On the radio, I discussed this with all of my squadron mates, and we decided I should disconnect and try a reconnection. So I did, and still, no fuel.

In the meantime, I was starting to sweat. So again, I tried several times to connect and disconnect with no success. Somehow, we had forgotten to monitor our alternates, and we had passed the point of no return. In other words, I didn't have enough gas to get to the alternate base, which, in this case, was Keflavik, Iceland. Despite this fact, my commander started to coordinate with the tanker for a diversion to the alternate. Because we were past the point of no return, this would mean a certain ditching in the Atlantic Ocean – I just didn't have enough fuel to make my alternate. Under our flight suits, we're required to wear an anti-exposure suit, which is basically a rubber suit, in case of a ditching, and I swear my boots were filling up with sweat as I contemplated my future. I determined that the cause for the malfunction was that the refueling system would not depressurize, which needs to happen before the tanker will initiate the fuel transfer. At this point, my commander finally made the call to leave formation with me on his wing and head toward the alternate, so I asked him if I could try one more refueling attempt except, this time, try a pressure refueling. This meant that I would hook up and the tanker would try to force the fuel into the aircraft against a pressurized intake nozzle. As I attempted this, with a deep sigh of relief, I watched the gas level begin to rise. For the rest of my life, I have reflected on this crisis as a near bailout and ditching. My own personal experience with a near ditching was very harrowing. Luckily, I didn't have to ditch in the water. However, there have been a number of aircraft ditchings that have been successful or nearly successful. The following[5] is a list of accidents that involved ditching an airliner:

> On 15 January 2009, US Airways Flight 1549 (an Airbus A320) successfully ditched into the Hudson River between New York City and New Jersey, after reports of multiple bird strikes. All of the 155 passengers and crew aboard escaped and were rescued by passenger ferries and day-cruise boats, in spite of freezing temperatures (the ditching occurred near the Circle Line Sightseeing Cruises and NY Waterway piers in midtown Manhattan). The survival rate was 100%.
>
> On 6 August 2005, Tuninter Flight 1153 (an ATR 72) ditched off the Sicilian coast after running out of fuel. Of 35 aboard, 23 survived with injuries, including serious burns. The plane's wreck was found in three pieces. The survival rate was 66%.
>
> On 16 January 2002, Garuda Indonesia Flight 421 (a Boeing 737) successfully ditched into the Bengawan Solo River near Yogyakarta, Java Island after experiencing a twin engine flameout during heavy precipitation and hail. The pilots tried to restart the engines several times before making the decision to ditch the aircraft. Of the 60 occupants, one flight attendant was killed. The survival rate was 98%. Photographs taken shortly after evacuation show that the plane came to rest in knee-deep water.
>
> On 23 November 1996, Ethiopian Airlines Flight 961 (a Boeing 767-260ER) ditched in the Indian Ocean near Comoros after being hijacked and running out of fuel, killing 125 of the 175 passengers and crew on board. Unable to operate flaps, it impacted at high speed, dragging its left

5 Retrieved from Wikipedia.org on Sept 2, 2009.

wingtip before tumbling and breaking into three pieces. The panicking hijackers were fighting the pilots for control of the plane at the time of the impact, which caused the plane to roll just before hitting the water, and the subsequent wingtip hitting the water and breakup are a result of this struggle in the cockpit. Some passengers were killed on impact or trapped in the cabin due to inflating their life vests before exiting. Most of the survivors were found hanging onto a section of the fuselage that remained floating. The survival rate was 29%.

On 2 May 1970, ALM Flight 980 (a McDonnell Douglas DC-9-33CF) ditched in mile-deep water after running out of fuel during multiple attempts to land at Princess Juliana International Airport on the island of Saint Maarten in the Netherlands Antilles under low-visibility weather. Of 63 occupants, 40 survivors were recovered by US military helicopters. The survival rate was 63%.

On 21 August 1963, an Aeroflot Tupolev Tu-124 ditched into the Neva River in Leningrad after running out of fuel. The aircraft floated and was towed to shore by a tugboat, which it had nearly hit as it came down on the water. The tug rushed to the floating aircraft and pulled it with its passengers near to the shore, where the passengers disembarked onto the tug; all 52 on board escaped without injuries. The survival rate was 100%.

On 28 September 1962, a Flying Tiger's Super H Constellation passenger aircraft with a crew of 8 civilian and 68 US military (paratrooper) passengers ditched in the North Atlantic about 500 miles west of Shannon, Ireland after losing three engines on a flight to Frankfurt, Germany. Forty-five of the passengers and 3 crew were rescued, with 23 passengers and 5 crewmembers being lost in the storm-swept seas. All passengers successfully evacuated the airplane. Those who were lost succumbed in the rough seas. The survival rate for landing and evacuation was 100%.

On 4 October 1960, 62 people died when Eastern Air Lines Flight 375 (Lockheed Electra 4-engine turbo-prop) plunged wing-first into Boston Harbor *after flying into a flock of starlings shortly after takeoff.* Three engines lost power, the plane stalled and spun, crashed into water 200 yards offshore, and broke in half. Nine of the 10 survivors had serious injuries. It was the first commercial airline crash in Logan Airport's history, the deadliest air disaster in New England history at the time, and it remains the most deadly crash in US history involving a bird strike. The survival rate was 14%.

In October 1956, Pan Am Flight 6 (a Boeing 377) ditched northeast of Hawaii, after losing 2 of its 4 engines. The aircraft was able to circle around USCGC Pontchartrain until daybreak, when it ditched; all 31 onboard survived. The survival rate was 100%.

In April 1956, Northwest Orient Airlines Flight 2 (also a Boeing 377) ditched into Puget Sound after what was later determined to have been caused by failure of the crew to close the cowl flaps on the plane's engines. All aboard escaped the aircraft after a textbook landing, but 4 passengers and 1 flight attendant succumbed either to drowning or to hypothermia before being recovered. The survival rate was 87%.

On 19 June 1954, Swissair Convair CV-240 HB-IRW ditched into the English Channel because of fuel starvation, which was attributed to pilot error. All 3 crew and 5 passengers survived the ditching and could escape the plane. However, 3 of the passengers could not swim and eventually drowned because there were no life jackets onboard, which was not prescribed at the time. The survival rate was 62%.

On 16 April 1952, the prototype de Havilland Australia DHA-3 Drover was ditched in the Bismarck Sea between Wewak and Manus Island. The port propeller failed, a propeller blade penetrated the fuselage, and the pilot was rendered unconscious; the ditching was performed by a passenger. The survival rate was 100%.

Pilots are not trained to ditch an airliner. They read about it in the regulations and are tested on it. But as you can see from the data above, there have only been 12 ditchings over the course of 57 years.

During the June 2009 NTSB hearings on Flight 1549, ditching became one of the main focus areas. The crew had done an outstanding job and the hearing focused on the amount and type of training the members of the aircrew had received. The aircraft, however, did not perform very well. The cargo bay was breached and forced up through the cabin floor, which seriously injured a flight attendant. The breach also allowed water to enter the cabin, causing the tail to drop and the nose to rise out of the water. The low attitude of the tail prevented the use of aft slide rafts. The following is an excerpt of the testimony that was presented before the Aviation Subcommittee on Transportation and Infrastructure, US House of Representatives by Captain John Prater, President of the Air Line Pilots Association (ALPA) on 23 February, 2009:

> The FAA requirements for designing airline aircraft are stringent, based on many years of manufacturing and operational experience, and are focused on providing the safest possible environment for occupants. Compliance with those standards makes it highly unlikely that serious problems will arise, but even the best-designed, best-operated aircraft cannot be made totally immune from danger. Thus, occupant survival provisions have been developed over the years which deal with everything from the flammability of fabric used for seat covers, to onboard medical care, to the possibility, however remote, of a ditching.
>
> Fortunately, because ditchings are such rare events, there is very little actual operational data we can examine to validate applicable design standards. The performance of the airplane structure during a ditching obviously cannot be tested during manufacturing flight test; it is done almost exclusively by analysis. The aviation community has been afforded an extraordinarily rare opportunity, therefore, the chance to analyze a virtually intact commercial airliner that

not only successfully landed on water, but also retained enough structural integrity to give all the occupants time to evacuate safely. We must, therefore, learn everything we can from this event, and the NTSB's investigation will lead the way to that knowledge.

We will not soon forget the dramatic footage of the survivors of Flight 1549 standing on the wings of the aircraft in life jackets and scrambling to reach life rafts. That scene highlighted two valuable safety features of airliners that may bear further investigation. The requirement for life jackets and other water-survival provisions is generally for those flights traveling some distance away from land. Indeed, some aircraft are designated as the "overwater" version of a common type to denote the fact that they are outfitted with those water survival provisions. However, some airlines are removing that equipment from those aircraft that do not fly extended distances from shore. ALPA suggests that it would be prudent for the FAA to revisit that practice and to undertake a detailed risk analysis to consider the possibility of a water landing in bodies of water other than the ocean, such as rivers and lakes.

Additionally, it appears that the emergency exits in the rear of Flight 1549 could not be used because that portion of the aircraft was partially submerged shortly after landing. The escape slides, which are required to be portable in order to be used at other exits, were therefore, unusable. It will be important as the investigation proceeds to determine why the aircraft floated in a tail-down position and what might be done to maximize the possibility that slide-rafts in all areas of the aircraft remain useable after ditching.

The passengers and crew of Flight 1549 were most fortunate in that the aircraft landed very near ferryboats, which enabled an almost-immediate rescue from the frigid waters of the Hudson. If the same successful ditching had occurred a few miles away, requiring passengers to board rafts and await rescue, it seems doubtful that sufficient raft space would have been available, and thus the outcome would not have been nearly so successful. ALPA urges the FAA, working with the NTSB investigation, to conduct a thorough analysis of the requirements for, and capabilities of, the various water survival provisions on aircraft used in commercial service.

On June 9, 2009, in response to the NTSB testimony, the FAA published the following fact sheet to clarify the FAA requirements for aircraft design and crew training related to ditching:

Federal Aviation Administration (FAA) rules to consider emergency landings on water – commonly known as "ditching" – have evolved over the last 50 years. Today, the regulations take into account the performance of both the airplane and crew, with the goal of providing the best possible chance for survival.

The Airplane

FAA certification regulations for transport category aircraft specify standards in two broad areas: the ditching characteristics of the plane and the ditching-related equipment the plane must carry. Section 25.801 of the regulations broadly states that the behavior of the airplane in a ditching situation must not cause immediate injury to the occupants or make it impossible for them to escape. The rules also say the flotation time and attitude of the airplane in the water must let the occupants evacuate the airplane.

The regulations (Section 25.807) also address the requirements for ditching exits. An airplane must have sufficient exits above the water level to allow passengers to evacuate before the sill of any ditching exit goes underwater. In reality, most airliners will continue to float beyond this time.

Sections 25.1411 and 1415 address requirements for ditching equipment, including the number of life rafts and their stowage. In general, there must be enough life rafts for all occupants even if the largest of the rafts is lost. The rules also specify that each occupant must have a life preserver readily available.

These regulations, which are usually associated with extended overwater flights, take into account preparations such as dumping fuel, closing valves and configuring the flight controls to enhance ditching behavior. They also consider the time needed to launch and board life rafts.

The Crew

Pilots must be familiar with ditching techniques, and ditching procedures exist in their aircraft manuals and checklists. Familiarization with the procedures is typically done in ground school. The FAA does not specifically require that pilots demonstrate ditching in a simulator.

Flight attendants are trained to respond to many different emergency situations, including ditching. Their training includes how to handle both anticipated and unplanned evacuations into water. They must be familiar with the life rafts, life preservers, flotation seat cushions, survival equipment and giving ditching instructions to passengers.

Ditching-related drills are required for flight attendants working for airlines with extended over-water flights. Training must include a planned ditching drill, preferably with the flight crew participating. They also must observe the deployment, inflation and detachment of each type of slide and life raft. The drills are required in both initial training and every 24 months thereafter.

Currently, the FAA does not *require* ditching drills be performed in water, but strongly recommends use of a realistic training environment. Another concern is whether the FAA and airlines need to revise emergency procedures for a double engine failure. For pilots, those procedures usually involve a checklist of many steps, and these checklists vary from problem to problem. If the plane is flying at a high altitude, then pilots may have time to identify and correct the problem, but at a low altitude, there's simply no time. Flight 1549's first officer, Jeffrey Skiles, said he only made it partly through the checklist for restarting the engines before the forced landing.

As an airline pilot, by FAA regulations, I am required to be familiar with the data in the airman's information manual (AIM) that describes ditching procedures,[6] many of which were extracted from the National Search and Rescue Handbook.[7]

Airline crewmembers are required by Federal Aviation Regulations[8] to receive training in certain emergency procedures. Each training program must provide emergency training for each aircraft type, model, configuration, and kind of operation conducted, as appropriate for each crewmember and certificate holder (e.g., pilot); emergency training must provide instruction in emergency assignments and procedures, including coordination among crewmembers; individual instruction in the location, function, and operation of emergency equipment, including equipment used in ditching and evacuation, first-aid equipment and its proper use, and portable fire extinguishers, with an emphasis on the type of extinguisher to be used on different classes of fires; instruction in the handling of emergency situations, including rapid decompression, fire in flight or on-the-surface, and smoke control procedures with an emphasis on electrical equipment and related circuit breakers found in cabin areas; ditching and evacuation; illness, injury, or other abnormal situations involving passengers or crewmembers; hijacking and other unusual situations; and finally, review of the certificate holder's previous aircraft accidents and incidents involving actual emergency situations.

Unless the Administrator finds that, for a particular drill, the crewmember can be adequately trained by demonstration, each crewmember must perform at least the following emergency drills,

6 See Appendix 5.

7 See Appendix 5.

8 According to Federal Aviation Regulations Part 135, Section 135.331.

using the proper emergency equipment and procedures: ditching, if applicable; emergency evacuation; fire extinguishing and smoke control; operation and use of emergency exits, including deployment and use of evacuation chutes, if applicable; use of crew and passenger oxygen; removal of life rafts from the aircraft, inflation of the life rafts, use of life lines, and boarding of passengers and crew, if applicable; and donning and inflation of life vests and the use of other individual flotation devices, if applicable.

Crewmembers who serve in operations above 25000 feet must receive instruction in the following: respiration; hypoxia; duration of consciousness without supplemental oxygen at altitude; gas expansion, gas bubble formation and physical phenomena and incidents of decompression.

On an annual basis or more frequently, I have received training in the procedures listed above. My airline requires me to pass an examination to demonstrate understanding. I have operated emergency doors, used aircraft fire extinguishers on simulated fires, practiced a simulated evacuation and slid down the slide. In the simulator, I have simulated a decompression and demonstrated the execution of an engine failure checklist from memory. In the airline pool, I have climbed into a raft that is installed on airliners and seen the emergency equipment demonstrated; I have donned a life vest, inflated it and stayed afloat in the pool while it was inflated.

Equipped with the above knowledge, I would handle a ditching that could occur during an overseas flight in the following manner. To aid in the rescue effort, if I had time and altitude (which Flight 1549 did not), I'd make a mayday call and transmit as many parameters as possible concerning the location and number of souls on board. I would turn on the emergency locator transmitter, which transmits a homing signal to rescue personnel through a satellite link. Then, I'd look down at the surface and try to find a ship that I could plan to ditch near. I wouldn't plan to ditch in front of the ship but, instead, use an offset so that I wouldn't be in line with the ship's path.

During the descent, I'd be trying to reduce aircraft weight by dumping fuel so that I could land in the water at the slowest possible speed. I'd plan on lowering the flaps to allow for further speed reduction and to allow the pitch attitude of the aircraft to remain as level as possible. I would make an announcement to the crew so that they could prepare the passengers for a water landing and evacuation. The passengers would put on their life preservers, tighten their seat belts and assume the brace position.

The success of the water landing would depend on the sea conditions, the wind and the skill of the pilot. I'd try to determine the direction of the wind by observing the swells in the water and plan to land upwind. An upwind landing would allow the aircraft to enter the water at a slower speed, thereby minimizing the damage. It is also important to determine which way the face of a swell is directed. Landing into the face of a swell could cause the aircraft to experience severe deceleration forces resulting in the break up of the aircraft, so I'd plan to land parallel to the swell.

At this point, I'd make an announcement to you, the passenger, to brace for impact. You might not know exactly what this means, but if you had read the card in the seat pocket in front of you, you would have found an explanation of this procedure. To reduce flailing upon impact, grab your ankles or legs or wrap your arms under your legs. Your head should face down in your lap and not be turned to either side. Feet should be flat on the floor and slightly ahead of the seat edge.

As I descend and line up for a landing, I would pick a path parallel to a swell and control my drift by taking a crab angle (i.e., a correction into the wind to avoid drift). If I were gliding with no engine power, I'd want to slow as much as possible, but I'd need to make sure that I maintained sufficient airspeed to avoid a stall. Above water, it is very difficult to judge your height as there is no depth perception, so it is very important to pay attention to the aircraft altimeter. Just before my water entry on landing, I'd want to slow the aircraft as much as possible and minimize the descent rate. If a crab angle was established, I would want to align the aircraft parallel to the swell by means of the rudder. It is important to enter the water in a level flight attitude. Depending on the location of the engines (most are under the wings), with the gear up, the engines may make first contact with the water. In the case of Flight 1549, one of the engines was sheared off during the water entry. It is important to keep the wings level because if they are not, one wing will most likely impact the water before the other, causing the aircraft to cartwheel and break apart. So, it's crucial to prepare for the fact that, once the aircraft enters the water, control will be lost.

No one can predict how long the aircraft will float in the water either, so it is best to evacuate the aircraft as soon as possible before it sinks. If you have your life preserver on, do not inflate it until outside the aircraft. In one of the accidents above, you may have noticed that some of the passengers were trapped in the cabin because they inflated their life preservers too soon. So, don't do that. The flight attendants will open the doors and deploy the life rafts. The life raft is attached to the aircraft and should be cut away as soon as everyone is in the raft slide, just in case the aircraft starts to sink.

Now that I've done my part as a pilot, it's your turn to do your part as a passenger – do your best to survive. This is your personal airliner ditching survival guide. Usually, the crew is responsible for leading the evacuation, survival and rescue efforts, but just to make sure that you're prepared for a worst-case scenario, let's assume no aircrew member survived the landing – you're left to fend for yourself. Hopefully, you paid attention during the flight attendant briefing on the ground and read the card in the seat pocket in front of you. So, what should you do now? Well, the most important things to remember are to exit the aircraft as soon as possible and try not to panic! Since you may only have a few minutes before the aircraft sinks, disconnect your seat belt. The life preserver is under the seat. If you have your life preserver on, don't inflate it as this could prevent you from leaving the aircraft. If you don't have it on, then grab it and take it with you. The seat cushion also floats, so you may want to take that with you too. Don't worry about your bags; just leave them because the evacuation will take place very quickly. Besides, the airline will most likely give you a significant bag reimbursement. If you are first to exit over the wing or out of a door, just open the door. In the case of the wing door, this is a plug-type hatch, and the emergency pull handle opens the door, which can then be removed by lifting it and pulling it inside the aircraft. All doors have slides that will automatically inflate when the door is opened. The doors are armed before takeoff and the slides should still be in the armed position. If the slides do not inflate automatically, then there is a manual inflation handle that can be pulled. The slides are configured to double as rafts and should be used in this manner.

If you are not the first out the door, then you need to look for a raft inside the aircraft. The location of the raft differs from aircraft to aircraft; however, in most airlines, decals indicate their location. You can always ask the flight attendant where the rafts are before takeoff. Some are stored in the overhead in a pull-down much like an attic door. Others are stored in a side panel. There are also first aid kits stored in various places, sometimes marked with decals, that can be taken. If water has entered the aircraft, the egress will be much more difficult. People who need wheelchairs, the elderly, and those who are hurt will need help, so be a good volunteer.

In the case of Flight 1549, not all of the slides deployed. The rear part of the aircraft was submerged, so there was no access to those slides. In a case like this, if you happen to be sitting in the back, you need to make new plans for leaving the aircraft. Because two of the slides were unusable, there wouldn't have been enough slides to contain all the passengers. If the aircraft were to sink, those who were not in slides would have to jump right into the water and, depending on which part of the planet you're in, the water may be very cold. Watch out because cold water will kill you. So, before you jump, make sure to inflate your life preserver. You can do this by blowing into the tube. Once you jump in, your body will go into a state of shock, and you will gasp for air. Your blood pressure and pulse rate will go up, so try not to panic. There will be a tendency to hyperventilate, so try to control your breathing. If possible, tread water or float on your back. The body loses heat in cold water 25 times faster than in air.

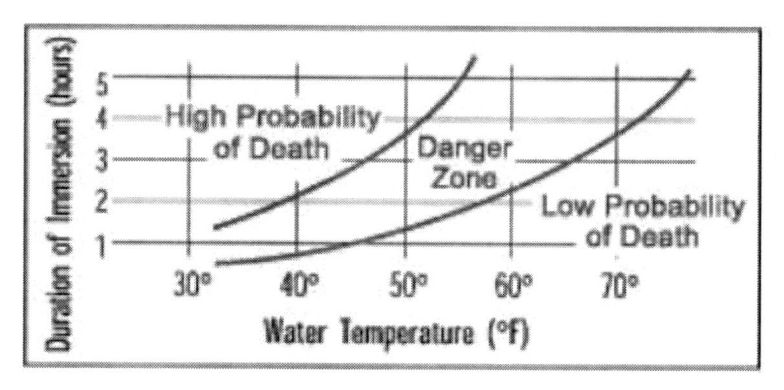

Figure 2.1 Hypothermia Chart[9]

Hypothermia Chart

IF THE WATER TEMPERATURE (F) IS:	EXHAUSTION OR UNCONSCIOUSNESS	EXPECTED TIME OF SURVIVAL IS:
32.5	Under 15 Minutes	Under 15 - 45 Minutes
32.5 - 40.0	15 - 30 Minutes	30 - 90 Minutes
40.0 - 50.0	30 - 60 Minutes	1 - 3 Hours
50.0 - 60.0	1 -2 Hours	1 - 6 Hours
60.0 - 70.0	2 - 7 Hours	2 - 40 Hours
70.0 - 80.0	3 - 12 Hours	3 Hours - Indefinitely
OVER 80.0	Indefinitely	Indefinitely

Table 2.1 Hypothermia Chart in Tabular Form

While you are in the water, your body will go through several stages. First, you will shiver and slur your words. Then, you will become exhausted and sleepy. Finally, your body will collapse, and you will become unconscious. The greatest threat to the passengers of Flight 1549 was hypothermia. The temperature of the Hudson River was 40 degrees Fahrenheit, and according to the table above, the passengers would have had 30 to 60 minutes before going unconscious.

If you're in an area where the surface is burning, take a deep breath and go underwater. Swim as far as you can before you come up to breathe. Before resurfacing, try to push the water aside. Go down into the water feet first and repeat the sequence until you are clear of the flames.

If you're lucky enough to be in the raft, make sure you wear your life vest at all times in case the raft capsizes. One person should tie

9 This chart was obtained from the US Coast Guard's Boating Safety Division and is shown in tabular form in Table 2.1.

themselves to the raft in case capsizing occurs. Make sure you enter the raft in the middle so that you don't fall out. Keep the weight in the raft evenly distributed. If you are the last one out of the aircraft, make sure that everyone has evacuated, and when the raft is full, cut it away from the aircraft. Look for evacuees in the water and toss them a line. Locate the slide survival kit and apply first aid to those that are injured. Designate a raft commander, and designate a buddy to each of the handicapped and elderly. Try to connect the rafts together but leave about 25 feet in between them to allow for wave motion. Finally, deploy the sea anchor, which acts to stabilize the raft.

Try to keep the floor of the raft dry, and if water enters the raft, bail it out. Install the raft canopy. The raft commander should collect drinking water and food and ration it. Tie the contents of the survival kit to the raft in case it capsizes. The contents of the survival kit vary from aircraft to aircraft, but FAA advisory circular 129-47 lists the minimum contents as follows: lines, including an inflation/mooring line with a snap hook, rescue or life line, and a heaving or trailing line; sea anchors; raft repair equipment (e.g., repair clamps, rubber plugs, and leak stoppers); inflation devices, including hand pumps and cylinders (i.e., carbon dioxide bottles) for emergency inflation; safety/inflation relief valves; canopy; hook-type knife, sheathed and secured by a retaining line; placards that give the location of raft equipment and are consistent with placard requirements; propelling devices (e.g., oars, or in smaller rafts, glove paddles); water catchment devices, including bailing buckets, reincatchment equipment, cups, and sponges; signaling devices, including at least one approved pyrotechnic signaling device, one signaling mirror, one spotlight or flashlight (including a spare bulb) having at least two 'V' cell batteries or equivalent, one police whistle, one dye marker, radio beacon with water-activated battery, radar reflector; one magnetic compass; a 2-day supply of emergency food rations supplying at least 1,000 calories a day for each person, one salt water desalting kit for each two persons the raft is rated to carry or two pints of water for each person the raft is rated to carry; one fishing kit; one book on survival, appropriate for any area and a survival kit, appropriately equipped. Other items which could be included in the survival kit are: triangular cloths, bandages, eye ointments, water disinfection tablets, sun protection balsam, heat retention foils, burning glass, seasickness tablets, ammonia inhalants and packets with plaster.

When I was a fighter pilot, I was continuously trained on the procedures for bailing out of an aircraft. You pull the ejection handle, the canopy blasts off, and you are fired out of the aircraft like a shotgun shell. As you drift down, the ejection seat falls away at a certain predetermined altitude; in other words, you are automatically separated from the seat. The seat of your pants, however, has a survival kit attached to it. So you basically sit on this survival kit every time you fly. It contains oxygen (so that you can breathe on your trip down from high altitudes), a one-person raft, and a personal survival kit. In addition to many of the same items listed above, the survival kit contains a GPS and beacon integrated into a survival radio.

While trying to survive in the water, you may encounter hazardous marine life, such as barracudas, moray eels, sea snakes, or sharks. I found information on sharing the ocean with these creatures in the Army Field Manual 21-76, Chapter 16, as well as on the Hawaiian Life Guard Association website. Barracudas are attracted to shiny objects and can bite, slash or tear your skin. Sea snakes are venomous but won't bite unless provoked. If you are bitten, you could suffer paralysis, severe pain or cardiac arrest. Moray eels won't bite unless provoked. A bite could cause severe muscle or tendon and nerve damage. Jellyfish have tentacles that will sting you and, in severe cases, a sting can cause difficulty with breathing and cardiac arrest. There are actually hundreds of species of sharks, but only 20 are known to attack. The most dangerous are the great white shark, the hammerhead, the mako, and the tiger shark. Other sharks known to attack man include the gray, blue, lemon, sand, nurse, bull, and oceanic white tip sharks. If you are in the water and a shark tries to attack, then yell and splash. If it attacks you, kick or strike the shark with an object. Hit the shark on the gills or eyes if possible. If you're in a raft and you spot a shark, don't let your arms or legs hang in the water. Keep quiet and don't move around. Don't fish; but if you do hook a fish, let it go. Don't throw garbage overboard. If you have to bury the dead at sea, do it at night. If a shark tries to attack, hit the shark with any object except your hand.

If someone becomes seasick, try to have them focus on the horizon. Vomiting can cause extreme fluid loss and dehydration and cause others to become seasick as well. To treat seasickness, wash the person and the raft to remove the sight and odor of vomit. There are seasickness pills in the survival kit, so use them. If someone develops saltwater sores, do not attempt to drain them. Flush the sores with fresh water and allow them to dry. Apply the antiseptic from the survival kit. If someone becomes blind or their eyes become sore from the sun, apply an ointment in the survival kit and bandage both eyes for 24 hours. If someone has severe sunburn, use the cream in the first aid kit. Try to stay in the shade and keep your head and skin covered.

Now that you've made it this far, assign someone to be the lookout for rescue aircraft or ships. If you see a rescue aircraft or ship, use items in the survival kit to signal. Have signal items available for instant use. When you are spotted and know you will be rescued, prepare for the rescue. Secure all loose items and fully inflate your life preserver if it has lost any air. When the rescue personnel arrive, stay in the raft and wait for instructions. The rescue aircraft will likely be a helicopter, in which case the rotor wash will be severe and may cause the raft to capsize. The helicopter will only be able to accommodate a limited number of people, so the raft commander should explain this to improve morale when the helicopter leaves.

So now you are a survivor, and by reading and rereading this personal airliner ditching survival guide, you will be better prepared to survive in the water. What you have just read may be your only source to successfully survive a water landing. I have crossed the Atlantic Ocean at least five hundred times and never thought it could happen to me, but pilots have to be fearless. The statistics from the ditchings described in this chapter are very favorable, so your chances of surviving a water landing are very good. As the saying goes, flying is basically hours of boredom punctuated occasionally by moments of shear terror. If those moments of terror happen to occur on an oceanic flight, you will now be prepared to survive.

Chapter 3:

FAA Aircraft Bird Strike Database

The following excerpt is taken from the Wright Brothers' diaries, 7 September 1905.

"Orville ... flew 4,751 meters in 4 minutes 45 seconds, four complete circles. Twice passed over fence into Beard's cornfield. Chased flock of birds for two rounds and killed one which fell on top of the upper surface and after a time fell off when swinging a sharp curve." This was the first reported aircraft bird strike. Judging from the closeness of the location to Dayton, Ohio and the time of year, the bird struck was probably a red-winged blackbird.

Since 1968, the FAA has requested that pilots report aircraft bird strikes, according to Sandra Wright and Dr. Richard Dolbeer from the USDA. In 1995, the USDA, Wildlife Services (USDA/WS) and National Wildlife Research Center took over management of the Wildlife Strike Database for the FAA through an interagency agreement. In 1999, Embry Riddle Aeronautical University (ERAU) took over management of the database. Then, in June 2009, ERAU created the International Center for Aviation and Wildlife Risk Mitigation to manage the bird strike database. Just two months after the Flight 1549 accident, though, the database was the subject of intense public debate as the FAA tried to shut down access. The reasons for this denial of access are explained below.

Pilots and other airport personnel are encouraged, but not required, to use the FAA form to report aircraft bird strikes. In March 2009, when the database was initially shut down by the FAA, the news media focused on the fact that reporting bird strikes was not required, and highlighted that as a major problem, Indeed, more data would enhance our understanding of the issue. The lack of a requirement to report was debated by the public, and no final resolution was reached. The more important point, though, is that the aircraft bird strike problem has been around since aircraft started flying 100 years ago. The NTSB identified aircraft bird strikes as a major problem by issuing their first safety recommendation in 1973. And yet, almost 40 years later, there is still no solution. In 2009, ALPA commented on this lack of progression and stated publicly that the FAA has been slow to fix this problem.

The background and method for reporting an airport wildlife strike is described in the FAA advisory circular (AC) 150/5200-32A, published in 2004. The following is an excerpt: "This Advisory Circular explains the importance of reporting collisions between aircraft and wildlife, more commonly referred to as 'wildlife strikes'. It also examines recent improvements in the Federal Aviation Administration's (FAA) Bird/Other Wildlife Strike Reporting system; how to report a wildlife strike; what happens to the wildlife strike report data; how to access the FAA National Wildlife Aircraft Strike Database; and the FAA's Feather Identification program."

Waterfowl (e.g., ducks and geese), gulls, and raptors (i.e., mainly hawks and vultures) are the bird species that cause the most damage to civil aircraft in the United States. And it is vultures and waterfowl that cause the most losses to US military aircraft in particular. To address this important safety issue, the FAA has initiated several programs, including the collection, analysis, and dissemination of wildlife strike data. Pilots, airport operations, aircraft maintenance personnel, or anyone who has knowledge of a strike is encouraged to report it to the FAA, as illustrated by the following instances: 1) a pilot reports striking one or more birds or other wildlife; 2) aircraft maintenance personnel identify aircraft damage as having been caused by a wildlife strike; 3) personnel on the ground report seeing an aircraft strike one or more birds or other wildlife; 4) bird or other wildlife remains, whether in whole or in part, are found within 200 feet of a runway centerline, unless another reason for the animal's death is identified; and 5) an animal's presence on the airport had a significant negative effect on a flight (i.e., aborted takeoff, aborted landing, high-speed emergency stop, aircraft left pavement area to avoid collision with animal).

The information for the rest of this chapter is derived from "Wildlife Strikes to Civil Aircraft in the United States 1990–2008", an FAA report authored by Sandra E. Wright, Dr. Richard A. Dolbeer, John Weller, and Michael J. Begier and published in September 2009. The FAA produced this report in cooperation with the US Department of Agriculture (USDA). Wildlife strikes may be reported to the FAA using a paper form[10] or electronically at the ERU Airport Wildlife Hazard Mitigation website.[11] The FAA's Bird/Other Wildlife Strike Report Form can be downloaded and printed from this website as well, and copies can also be found in the Airman's Information Manual (AIM).

The FAA National Wildlife Strike Database Manager edits all strike reports to ensure consistent, error-free data before entering the report into the database. This information is supplemented with non-duplicated strike reports from other sources. About every six weeks, an updated version of the database is posted online. In addition, once a year, a current version of the database is forwarded to the International Civil Aviation Organization (ICAO) for incorporation into ICAO's Bird Strike Information System Database.

10 FAA Form 5200-7 Bird/Other Wildlife Strike Report. Paper copies of Form 5200-7 may also be obtained from the appropriate Airports District Offices (ADO), Flight Standards District Offices (FSDO), and Flight Service Stations (FSS). Copies of the Bird/Other Wildlife Strike Report form are also found in the Airman's Information Manual (AIM).

11 http://wildlife.pr.erau.edu/public/index.html

Analyses of data from the FAA National Wildlife Aircraft Strike Database has proved invaluable in determining the nature and severity of the wildlife strike problem. The database provides a scientific basis for identifying risk factors; justifying, implementing and defending corrective actions at airports; and for assessing the effectiveness of those corrective actions. The database is invaluable to engine manufacturers and aeronautical engineers, in particular, as they develop new technologies for the aviation industry. Each wildlife strike report contributes to the accuracy of, and effectiveness of, the database. Moreover, each report contributes to the common goal of increasing aviation safety.

In order to access the FAA National Wildlife Aircraft Strike Database and expedite the dissemination of this important information, the FAA has developed procedures for searching the database online. The public may access the database without a password and retrieve basic information on the number of strikes by year, by state, and by species of wildlife. Database access for airport operators, airline operators, engine manufacturers, air frame manufacturers, and certain other governmental agencies requires a password and allows retrieval of more detailed wildlife strike information for their specific area of concern. Airport operators' access is limited to strike information for incidents occurring on the grounds of their particular airport. Airlines may only access strike records involving aircraft owned or operated by them. Thus, comparisons among individual airports and airlines are not made. To gain access to the FAA National Wildlife Aircraft Strike Database, airline and airport operators, airframe and engine manufactures, and governmental agencies must write to the FAA Staff Wildlife Biologist.

Accurate species identification is critical for aircraft bird strike reduction programs. In order to make proper management decisions, wildlife biologists must know what species of animal they are dealing with. The FAA, USAF, USDA/WS are working closely with the Feather Identification Lab at the Smithsonian Institution (SI) at the Museum of Natural History, to further our understanding and prevention of bird-aircraft strike hazards. Bird strike remains that cannot be identified by airport personnel or by a local biologist can be sent (with FAA Form 5200-7) to the Smithsonian Museum for identification. Feather identification of birds involved in bird-aircraft strikes will be provided free of charge to all US airport operators, all US aircraft owners/operators (regardless of where the strike happened), or to any foreign air carrier if the strike occurred at a US airport. The following guidelines need to be observed for collecting and submitting feathers or other bird/wildlife remains for species identification. These guidelines help maintain species identification accuracy, reduce turn-around time, and maintain a comprehensive FAA National Wildlife Aircraft Strike Database. Collect and submit remains as soon as possible. Provide complete information[12] regarding the incident.

Mail the report with the feather material. Provide your contact information if you wish to be informed of the species identification. Mail the Bird/Other Wildlife Strike Report and collected material to the Smithsonian's Feather Identification Lab.[13] They will forward the report to the FAA Staff Wildlife Biologist at the FAA's Office of Airport Safety and Standards. This service is provided free of charge to all US-registered aircraft owners/operators, regardless of where the strike occurred, and to all aircraft owners/operators when the strike occurred at a US airport.

An article titled "Species Identification of Bird Strike Remains" by Marcy Heacker and Carla J. Dove at the SI Feather Lab explains the identification process in more detail. The following is an excerpt from that article:

> The SI Feather Identification Lab is a highly specialized lab that currently processes over 3,000 cases a year for bird species identification from whole and fragmentary feather

12 Fill out FAA Form 5200-7 – Bird/Other Wildlife Strike Report, a copy of which can be downloaded and or printed from: http://wildlife.pr.erau.edu/strikeform/birdstrikeform.php

13 The address is: Feather Identification Lab, Smithsonian Institution, NHB, E600, MRC 116, P.O. Box 37012, Washington, D.C. 20013-7012.

material. The lab is housed in one of the world's best museum collections of birds at the National Museum of Natural History in Washington, DC. This comprehensive collection is ideal for the work done today by researchers Carla Dove (Program Manager), Marcy Heacker (Research Assistant), and Nancy Rotzel (Molecular Specialist).

While each case is different, the approach to the identification of bird strike remains is based on what kind of material is available. If there is a whole bird or partial carcass, identifications can be based on physical characteristics traditionally used when viewing birds in the wild – including size, color, and pattern. Wings, feathers, feet, and beaks can then be compared with the bird specimens in the SI museum collection to make a final identification. This approach is also applied when samples only include loose or fragmented feathers.

Often, there is very little material recovered from a bird strike. Identification of samples consisting of small feather fragments, blood, and/or tissue can be examined in a couple of ways. The microscopic features of the downy part of a feather are unique for different groups of birds (e.g., duck, raptor, or passerine). Looking at this fluffy area of the feather can provide valuable clues to narrowing down the species identification.

The latest tool in the Feather Lab's identification toolbox is DNA analysis. Using molecular techniques to analyze minute bird strike remains is an important new advancement in the lab's ability to identify birds from blood and tissue samples. Because the type of remains is never the same, the best way to approach collecting bird strike remains depends on what kind of material is available. The following is a breakdown of how to collect material for species identification for each of these types of remains.

Whole/Partial Bird: Many times, a whole bird is found on

the airfield. In these cases, pluck a variety of feathers (breast, back, wing, and tail) and place in a re-sealable plastic bag. The remaining carcass can be properly discarded. Feet, whole wings, or beaks can also be helpful if they can be removed from the whole carcass. If only part of the bird is available, collect the best variety of feathers possible; particularly feathers with obvious color or pattern. If the remains are moist or fleshy, allow them to dry and wrap them in paper before placing in re-sealable plastic bag and/or double-bag them.

If only feather fragments, tissue or blood is available, the best thing to do for a blood smear or tissue ("snarge") is to send all of the material. Paper towels, gauze pads, alcohol wipes work well for collecting this type of material. If the area is dry, it is best to moisten the material with a spray of 70% alcohol. Again, place everything in a re-sealable plastic bag to ship to the lab. The SI DNA work has found that dry material or material soaked in alcohol gives the best results. Don't use water as it encourages nasty molds and funguses that degrade DNA.

Now that you understand the reporting and identification process, it's time to move on to the current details of the database. Since 1995, the FAA has produced annual reports. An analysis of the most recent wildlife strike databases was published[14] in September 2009 and covers the 19-year period from 1990 to 2008. The following are excerpts from this report:

> Many populations of wildlife species commonly involved in strikes have increased markedly in the last few decades and adapted to living in urban environments, including airports. For example, from 1980 to 2007, the resident (non-migratory) Canada goose population in the USA and Canada increased at a mean rate of 7.3% per year. Other species showing significant mean annual rates of increase included bald eagles (4.6%); wild turkeys (12.1%); turkey vultures (2.2%), American white pelicans (2.9%), double-crested cormorants

14 with serial report number 15

(4.0%), and sandhill cranes (5.0%). Thirteen of the 14 bird species in North America with mean body masses greater than 8 lbs. have shown significant population increases over the past three decades (Dolbeer & Eschenfelder 2003). The white-tailed deer population increased from a low of about 350,000 in 1900 to over 30 million in the past decade (McCabe & McCabe 1997, Hubbard *et al.* 2000, Adams *et al.* 2005).

Concurrent with population increases of many large bird species, air traffic has increased substantially since 1980. Passenger enplanements in the USA increased from about 310 million in 1980 to a record 750 million in 2008 (3.2% per year), and commercial air traffic increased from about 18 million aircraft movements in 1980 to over 28 million in 2008 (1.6% per year, Federal Aviation Administration 2009). USA commercial air traffic is predicted to continue growing at a rate of about 1.3% per year to over 35 million movements by 2025.

Commercial air carriers are replacing their older three- or four-engine aircraft fleets with more efficient and quieter, two-engine aircraft. In 1965, about 90% of the 2,100 USA passenger aircraft had three or four engines. In 2005, the USA passenger fleet had grown to about 8,200 aircraft, and only about 10% have three or four engines (Department of Transportation 2009). This reduction in engine redundancy increases the probability of life-threatening situations resulting from aircraft collisions with wildlife, especially with flocks of birds. In addition, previous research has indicated that birds are less able to detect and avoid modern jet aircraft with quieter turbofan engines (Chapter 3, International Civil Aviation Organization 1993) than older aircraft with noisier engines (Burger 1983, Kelly *et al.* 1999).

As a result of these factors, experts within the Federal Aviation Administration (FAA), U.S. Department of Agriculture (USDA), and U.S. Air Force expect the risk, frequency, and potential severity of wildlife-aircraft collisions to grow over the next decade. The FAA has initiated several programs to address this important safety issue. Among the various programs is the collection and analysis of data from wildlife strikes. The FAA began collecting wildlife strike data in 1965. However, except for cursory examinations of the strike reports to determine general trends, the data were never submitted to rigorous analysis. In 1995, the FAA, through an interagency agreement with the USDA, Wildlife Services, (USDA/WS), initiated a project to obtain more objective estimates of the magnitude and nature of the national wildlife strike problem for civil aviation. This project involves having specialists from the USDA/WS: (1) edit all strike reports (FAA Form 5200-7, *Birds/Other Wildlife Strike Report*) received by the FAA since 1990 to ensure consistent, error-free data; (2) enter all edited strike reports in the FAA National Wildlife Strike Database; (3) supplement FAA-reported strikes with additional, non-duplicated strike reports from other sources; (4) provide the FAA with an updated computer file each month containing all edited strike reports; and (5) assist the FAA with the production of annual and special reports summarizing the results of analyses of the data from the National Wildlife Strike Database. Such analyses are critical to determining the economic cost of wildlife strikes, the magnitude of safety issues, and most important, the nature of the problems (e.g., wildlife species involved, types of damage, height and phase of flight during which strikes occur, and seasonal patterns). The information obtained from these analyses provides the foundation for refinements in the development, implementation, and justification of integrated research and management efforts to reduce wildlife strikes.

For the 18-year period (1990-2008), 89,727 strikes were reported to the FAA. Birds were involved in 97.4% of the reported strikes, terrestrial mammals in 2.1%, bats in 0.3% and reptiles in 0.1%. The number of strikes annually reported more than quadrupled from 1,759 in 1990 to a record 7,516 in 2008. We suggest that the increase in reports from 1990 to 2008 was the result of several factors: an increased awareness of the wildlife strike issue, an increase in aircraft operations, an increase in populations of hazardous wildlife species, and an increase in the number of strikes (Dolbeer 2000, Dolbeer & Eschenfelder 2003).

Most (66%) of the 89,727 strike reports were filed using the paper (43%) or electronic (23%) version of FAA Form 5200-7, *Bird / Other Wildlife Strike Report*. Since the online version of this form became available in April 2001, use of the electronic reporting system has climbed dramatically. In 2008, 68% of the strike reports were submitted electronically. Airline personnel and pilots filed 29% and 24% of the strike reports, respectively. About 85% of the reported strikes involved commercial aircraft; the remainder involved business, private, and government aircraft Reports were received from all 50 states, from some USA territories, and from foreign countries when USA-registered aircraft were involved California, Texas, Florida, and New York had the most (7,442, 5,963, 5,571, and 4,732, respectively) bird strike reports. Twenty-one other states each had more than 1,000 bird strikes reported. New York, California, Illinois, New Jersey, Texas, Colorado, and Michigan each had 100 or more terrestrial mammal strikes. In all, strikes were reported at 1,625 airports (1,456 airports in the USA and 215 foreign airports where USA-based aircraft were involved).

Most bird strikes (51%) occurred between July and October; 62% occurred during the day; 60% occurred during the landing (descent, approach, or landing roll) phase of flight; and 37% occurred during takeoff and climb. Most terrestrial mammal strikes (56%) occurred between July and November with 32% of deer strikes concentrated in October- November. Most terrestrial mammal strikes (64%) occurred at night, 55% occurred during the landing roll, and 34% occurred during the takeoff run.

About 59% of the bird strikes occurred when the aircraft was at a height of 100 feet or less AGL, 72% occurred at 500 feet or less AGL, and 92% occurred at or below 3,000 feet AGL.

Less than 2% of bird strikes occurred above 10,000 feet AGL. The record height for a reported bird strike involving civil aircraft in USA was 32,500 feet AGL. Terrestrial mammal strikes predominately occurred at 0 feet AGL; however, 9% of the reported strikes occurred while the aircraft was in the air, e.g., when the aircraft struck deer with the landing gear.

The aircraft components most commonly reported as struck by birds were the nose/radome, windshield, engine, wing/rotor, and fuselage. Aircraft engines were the component most frequently reported as being damaged by bird strikes (32% of all damaged components). There were 11,060 strike events in which a total of 11,616 engines were reported as struck (10,525 events with one engine struck, 518 with two engines struck, 12 with three engines struck and 5 with four engines struck (9,877 events with one engine struck, 986 with two engines struck, 33 with three engines struck, and 20 with four engines struck). In 3,484 damaging bird strike events involving engines, a total of 3,596 engines were damaged (3,375 events with one engine damaged, 107 with two engines damaged, 1 with three engines damaged and 1 with 4 engines damaged). Aircraft components most commonly reported as struck by terrestrial mammals were the landing gear, propeller, and wing/rotor. These same components ranked highest for the parts most often reported as damaged by mammals.

Of the 87,416 bird strikes reported, 68,653 provided some indication as to the nature and extent of any damage. Of these 68,653 reports, 59,047 (86%) indicated the strike did not damage the aircraft; 5,112 (7%) indicated the aircraft suffered minor damage; 2,456 (4%) indicated the aircraft suffered substantial damage; 2,015 (3%) reported an uncertain level of damage; and 24 reports (less than 1%) indicated the aircraft was destroyed as a result of the strike.

Of the 1,912 terrestrial mammal strikes reported, 1,246 reports provided some indication as to the nature and extent of any damage. Of these 1,246 reports, 508 (41%) indicated the strike did not damage the aircraft; 324 (26%) indicated the aircraft suffered minor damage; 331 (27%) indicated the aircraft suffered substantial damage; 58 (5%) reported an uncertain level of damage; and 25 (2%) indicated the aircraft was destroyed as a result of the strike.

Not surprisingly, a much higher percentage of terrestrial mammal strikes (59%) resulted in aircraft damage than did bird strikes (14%). Deer (782 strikes) were involved in 41% of the 1,912 terrestrial mammal strikes. In 13% and 54% of the bird and terrestrial mammal strike reports, respectively, an adverse effect-on-flight was reported. Three percent of bird strikes resulted in an aborted takeoff compared to 17% of terrestrial mammal strikes.

Only 38,474 (44%) of the 87,416 bird strike reports provided information on the type of bird (e.g., gull or hawk). Furthermore, only 24,351 (63%) of these 38,474 reports provided identification to species level (e.g., ring-billed gull or red-tailed hawk). Thus, birds were identified to the species level in only 28% of the 87,416 reported bird strikes. In all, 381 identified species of birds were struck; 176 identified species were reported as causing damage. Gulls (19%), doves/pigeons (15%), raptors (13%), and waterfowl (8%) were the most frequently struck bird groups. Gulls were involved in 2.4 times more strikes than waterfowl (7,470 and 3,175, respectively). Waterfowl, however, were involved in 1.2 times more damaging strikes (1,418 or 31% of all damaging strikes in which the bird type was identified) than were gulls (1,169 or 25% of all damaging strikes in which the bird type was identified). Gulls were responsible for the greatest number of bird strikes (935 or 27%) that had a negative effect-on-flight. The most frequently struck terrestrial mammals were Artiodactyls – primarily deer (43%) – and carnivores – primarily coyotes (34%). Artiodactyls were responsible for 92% of the mammal strikes that resulted in damage and 79% of the mammal strikes that had a negative effect-on-flight. In all, 33 identified species of terrestrial mammals and 8 identified species of bats were reported struck; 19 identified species of terrestrial mammals and 1 identified species of bat caused damage.

For the 19-year period, reports were received of 9 wildlife strikes that resulted in 16 human fatalities. Five of these strikes resulting in 7 fatalities involved unidentified species of birds. Canada geese, white-tailed deer and brown-pelicans were responsible for the other 9 fatalities. Reports were received of 167 strikes that resulted in 209 human injuries. Waterfowl (ducks and geese; 40 strikes, 45 humans injured), vultures (24 strikes, 26 injuries), and deer (18 strikes, 25 injuries) caused 82 of the 132 strikes resulting in injuries in which the species or species group was identified.

For the 19-year period, reports were received of 49 aircraft destroyed or damaged beyond repair due to wildlife strikes. The majority (63%) were small (<2,250 kg maximum takeoff mass) general aviation (GA) aircraft. Terrestrial mammals (primarily white-tailed deer) were responsible for 25 (51%) of the incidents. Canada geese (4 incidents) and vultures (3 incidents) were each responsible for 7 (60%) of the 14 incidents involving birds in which the species or species group was identified. Thirty-three (67%) of the 49 wildlife strikes resulting in a destroyed aircraft occurred at GA airports, 9 occurred away from an airport, and 6 occurred at airports certificated for passenger service under 14 CFR Part 139. GA airports, often located in rural areas with inadequate fencing to exclude large mammals, face unique challenges in mitigating wildlife risks to aviation (DeVault *et al.* 2008, Dolbeer *et al.* 2008).

For the 19-year period, reported losses from bird strikes totaled 393,521 hours of aircraft downtime and $308.3 million in monetary losses. Reported losses from terrestrial mammal strikes totaled 244,068 hours of aircraft downtime and $38.8 million in monetary losses. Bat strikes resulted in 100 hours of aircraft downtime and $3.2 million in losses. Reptile strikes resulted in 3 hours of aircraft downtime. Of the 15,179 reports that indicated the strike had an adverse effect on the aircraft and/or flight, 4,301 provided an estimate of the aircraft down time (Σ = 637,692 hours, avg. = 148.3 hours down time/incident). Of the reports providing a damage cost estimate for the incident, 2,620 gave an estimate of the direct aircraft damage cost (Σ = $308.6 million, avg. = $11,787 damage/incident), and 1,157 gave an estimate of other monetary losses (Σ = $41.1 million, avg. = $36,003 lost/incident). Other monetary losses include such expenses as lost revenue, the cost of putting up passengers in hotels, re-scheduling aircraft, and flight cancellations. Analysis of strike reports from USA airports and airlines indicated that less than 20% of all strikes were reported to the FAA (Cleary *et al.* 2005, Wright & Dolbeer 2005).

Additionally, only 28% of the 15,179 reports indicating an adverse effect provided estimates of aircraft downtime, 17% provided estimates of direct costs, and 8% provided estimates of other (indirect) costs. Furthermore, many reports providing cost estimates were filed before aircraft damage and downtime had been fully assessed. As a result, the information on the number of strikes and associated costs compiled from the voluntary reporting program is believed to severely underestimate the magnitude of the problem.

Assuming (1) all 15,179 reported wildlife strikes that had an adverse effect on the aircraft and/or flight engendered similar amounts of downtime and/or monetary losses and (2) that these reports are all of the damaging strikes that occurred, then, at a minimum, wildlife strikes cost the USA civil aviation industry 118,448 hours per year of aircraft downtime and $123 million in monetary losses ($94 million per year in direct costs and $29 million per year in associated costs). Further, assuming a 20 percent reporting rate, the annual cost of wildlife strikes to the USA civil aviation industry is estimated to be in excess of 592,000 hours of aircraft downtime and $614 million in monetary losses ($470 million per year in direct costs and $144 million per year in associated costs).

An analysis of 19 years of strike data reveals the magnitude and severity of the wildlife-aircraft strike problem for civil aviation in the USA. Wildlife strikes continue to pose a significant economic and safety risk for civil aviation in the USA. Management actions to reduce wildlife strikes are being implemented at many airports (*e.g.*, Wenning *et al.* 2004, DeFusco *et al.* 2005, Dolbeer 2006a, Human Wildlife Conflicts Journal 2009), and these efforts may be responsible for the general decline is reported strikes with damage from 2000-2008. To address the problem, airport managers first need to assess the wildlife hazards on their airports with the help of qualified airport biologists (FAA Advisory Circular 150/5200-36). They then must take appropriate actions, under the guidance of professional biologists trained in wildlife damage management, to minimize the risks posed by wildlife. The aviation community must also widen its view of wildlife management to consider habitats and land uses in proximity to the airport. Wetlands, dredge spoil containment areas, waste disposal facilities, and wildlife refuges can attract hazardous wildlife. Such land uses are often incompatible with aviation safety and should either be prohibited near airports or designed and operated in a manner that minimizes the attraction of hazardous wildlife. The manual *Wildlife Hazard Management at Airports* (Cleary & Dolbeer 2005) provides guidance to airport personnel and biologists for conducting wildlife hazard assessments (WHA) an in developing and implementing wildlife hazard management plans.

Finally, there is a need for increased and more detailed reporting of wildlife strikes. Previous analyses (Cleary *et al.* 2005, Wright & Dolbeer 2005) indicated that less than 20% of all wildlife strikes involving USA civil aircraft are reported; new information indicates that approximately 39% of wildlife strikes are now reported (Dolbeer unpublished data 2009). Although the quantity of strike reporting is higher more detail is needed. Approximately 44% of all reported bird strikes for 1990-2008, provided information on the type of bird struck, and only about 28% of the reports identified the birds struck to species level. In addition, only 17% of strike reports indicating an adverse effect on the aircraft or flight provided at least a partial estimate of economic losses resulting from the strike.

As previously mentioned, the FAA wildlife database became the subject of recent congressional hearings on US Airways Flight 1549 in February 2009. The public wanted reporting of bird strikes to be made mandatory, and the FAA's testimony told why reporting is hard to enforce. The following are excerpts from that hearing:

> Currently, the database has 106,604 records from January 1990 through August 2008. The increasing number of bird strikes is a combination of better reporting and increasing bird populations. The database is available to airport operators and safety analysts and is extremely useful for determining which species are most frequently involved in strikes, seasonal patterns, and extent and type of damage from strikes.
>
> Mandatory reporting of wildlife strikes is extremely difficult to enforce and may not necessarily increase accurate reporting. The success of the voluntary reporting system is proven by the increase in annual reports from only 1,900 reported strikes in 1990, to almost 8,000 reported strikes in 2007. Advances in wildlife strike reporting through Web-based technology make it easier and faster to report strikes. Moreover, the FAA, in close partnership with the USDA, continues to educate and increase awareness through 3 ongoing campaigns in concert with industry, conferences and participation on the National Bird Strike Committee.
>
> Our statistics on bird strikes indicate that the closer the aircraft is to the runway, the higher the risk of a bird strike. Conversely, the risk of a substantial bird strike decreases significantly with altitude. High altitude strikes are not common, though they do occur. For instance, at 30,000 feet, there was only one reported bird strike between 1990-2008. However, about 73% of all strikes occur within the airport environment up to 500 feet above ground. According to reports, Flight 1549

had reached an altitude of 3,200 feet when it encountered a flock of Canada geese that resulted in numerous bird strikes to the airframe and engines. Since the data indicate that the greatest risk of bird strikes occurs at the airport, the FAA has focused its bird strike mitigation efforts at airports. By regulation, the FAA requires commercial service airports to maintain a safe operation. This includes conducting Wildlife Hazard Assessments (WHA) and preparing a Wildlife Hazard Management Plan, if necessary.

On March 19th, 2009 the FAA's Acting Associate Administrator for Airports released a proposed rule change which would protect information in the FAA Wildlife Hazard Database from public disclosure. The FAA was "concerned that there is a serious potential that information related to bird strikes will not be submitted because of fear that the disclosure of raw data could unfairly cast unfounded aspersion on the submitter." Therefore, bird strike reporting would decrease. Further, the FAA stated that "When the FAA began collecting this data; it assured the entities submitting the data that the submissions would not be made available to the public." In fact, database information had been released a number of times, including to the press.

The proposed rule entered a 30-day public comment period. During this period, over 55 comments were received at the www.regulations.gov website. The vast majority were from the general public who uniformly opposed the FAA action. While there were a few comments from industry supporting the FAA's proposal, public comments ran about 5-1 against it. Most commenters felt the FAA was attempting to "hide" something.

The National Transportation Safety Board, which investigates airline and other transportation accidents, told the agency in a letter that a lack of public information "could hamper efforts to understand the nature and potential effects of wildlife threats to aviation." The safety board believes mandatory reporting of all bird strikes would allow more complete and accurate assessment of the problem.

On April 21, the day after the public comment period closed, Transportation Secretary La-Hood told the *Washington Post*: "I think all of this information ought to be made public. . . . We're going to, you know, make this information as public as anybody wants it." Further, La-Hood stated that the FAA's efforts to keep bird strike data secret "...doesn't comport with the President's idea of transparency." He also stated that "...it's something that somebody wanted to put out there to get a reaction. We got the reaction, and now we're going to bring it to conclusion."

On April 22, the FAA issued a press release in which they announced that not only would the database information be available to the general public but that the database would be posted on the FAA Wildlife Mitigation website. According to the press release, "...the FAA has determined that it can release the data without jeopardizing aviation safety. . . . over the next four months, the FAA will improve the search function and make it more user-friendly."

The current totals in the database, as of the publishing of this book, for the period January 1, 1990 through September 30, 2009 are 101,198 bird strikes with 91,399 reported at civil airports and 9,799 at joint-use military/civil airports. There was also a total of 461 species in the database. At New York's LaGuardia Airport, where Capt. Sullenberger's US Airways flight originated, the total number of wildlife strikes for Jan 1990 thru July 2009 was 952. For Jan-July 2009, the number of reported strikes was 15. Since only 20% of bird strikes are reported, though, the number of actual bird strikes is estimated to be 75, which is 2 strikes per week.

CHAPTER 4:

Aircraft Accident Investigations

So after the airliner ditched in the Hudson, who was notified? How did the accident investigation process proceed? Well, it may seem to the public like a chaotic process because they hear about it from so many organizations through the news media; however, there is a very detailed and well-organized process that is followed strictly as described in the NTSB Aviation Investigation Manual, the Flight Data Recorder Handbook, the Cockpit Voice Recorder Handbook and the Methodology for Investigating Operator Fatigue in a Transportation Accident.

The airport first notifies the NTSB Communication Center, which is located in Washington DC. The NTSB is responsible for investigating every aviation accident in the United States. The NTSB investigates and reports on all US air carrier accidents, commuter and air taxi crashes, mid-air collisions, serious mishaps involving public-use (government) aircraft and all fatal general aviation accidents. The NTSB also investigates accidents involving both civilian and military aircraft as well as crashes involving military aircraft where the functions of the FAA are at issue. Internationally, the NTSB investigates major accidents involving US air carriers and US-manufactured airliners under the auspices of the International Civil Aviation Organization (ICAO).

On some occasions, the NTSB may delegate the investigation to the FAA. This will occur if the accident is a non-fatal, general aviation accident, involving fixed-wing aircraft of less than 12,500 pounds, home-built aircraft, crop dusters or rotorcraft; but the NTSB retains the discretion to oversee these accidents, if necessary. Also worth noting, when FAA investigators are working for the NTSB, they are not supposed to use information acquired during the NTSB accident investigation for FAA License Enforcement purposes.

The NTSB Communications Center then notifies the Go Team, whose purpose is to begin the investigation of major accidents at the accident scene, as quickly as possible, assembling a broad spectrum of technical expertise to solve complex transportation safety problems. The Go Team can number from three or four to more than a dozen specialists from the Board's Headquarters staff in Washington, D.C. who are assigned on a rotational basis to respond as quickly as possible to the scene of the accident. Go Teams travel by commercial airliner or government aircraft, depending on circumstances and availability. Such teams have been flying to catastrophic airline crash sites for more than 35 years. They also routinely handle investigations of certain rail, highway, marine and pipeline accidents. A seasoned NTSB member who is called the Investigator in Charge (IIC) is assigned to lead the Go Team. The IIC divides the investigation team into working groups, each with its own specialty and staff. A description of each group and their functions is as follows:

The Air Traffic Control Group gathers the transcripts in the control tower. They also obtain the recorded radar data from the subject aircraft. The operations group investigates the pilot's recent duty history. The meteorology group gathers data from the National Weather Service and looks at the aircraft log book to examine weather conditions both at and around the accident scene. The systems group examines the aircraft electrical, hydraulic, pneumatic, flight controls, navigation and warning systems. The power plants group analyzes the engines. The maintenance group obtains information on the aircraft and its history to determine if any previous related malfunctions could have contributed to the accident. The records group looks at the pilots' training and testing records. The survival factors group examines evacuation and rescue efforts. The aircraft performance group investigates the flight parameters of the aircraft from before, during and after the accident. The human performance group looks at human influences that occurred before the accident, such as fatigue, medication and alcohol. The structures group takes pictures of the aircraft wreckage and accident scene. The cockpit voice recorder (CVR) group examines the conversations between the pilots to determine if correct operational procedures were followed. The flight data recorder (FDR) group investigates the flight instrument recordings to reconstruct the flight. Many times, this information is used in a simulator to provide a visual display of the aircraft before, during and after the accident. Each group chairman prepares a report and submits it to the IIC.

The composition of the Go Team for a specific accident will depend on the number of injuries/fatalities, type of aircraft, previous 5 accidents of this type, location of the accident, extent of aircraft or ground damage, weather, public interest, and specialist workloads. The Go Team is on-call 24 hours per day, 365 days per year, and is required to leave within two hours of being notified. One of the members will leave immediately to perform public affairs functions, such as notifying the news media that the investigation is under NTSB jurisdiction.

The working groups may remain at the accident scene anywhere from several days to several weeks. The US Go Team only responds to accidents in the United States, so if a US-registered aircraft has an accident in another country, then that country is normally responsible for the investigation. Many times, though, an IIC from the NTSB will assist in that sort of investigation.

In addition to the Go Team, the NTSB invites the FAA, pilots, controllers, aircraft manufacturers, engine manufacturers, and labor organizations to be members of the investigative team. The air carrier, airframe manufacturer and the FAA are almost always part of the team. The police, firefighters, FBI, FEMA and other agencies can assist too, but they are not members of the investigation team.

After the IIC arrives at the accident scene, he sets up a command post, often in a nearby hotel. The IIC will also arrange for a room to use for press briefings. Only confirmed, factual information is released to the media, so there is no speculation over the cause of an accident. At this point, the first meeting is held and is considered to be the formal opening of the investigation. At this meeting, the facts of the accident will be presented and may include the name of the pilot, aircraft type and registration number, type of flight, origin and intended destination, number of fatalities (or best information presently known), condition and location of crewmembers, extent of aircraft damage; and other information considered relevant (e.g., hazardous material (HAZMAT) and site considerations).

Next, the IIC will make an opening statement that may include mention of the following: safety board authority to conduct the investigation; the roles of the board members, roles of parties to the investigation, roles of international participants in the investigation (i.e., accredited representatives and technical advisors); the organization of the team into groups of specialists; qualifications of personnel participating in the investigation; expected participation of participants for the duration of on-site activities and follow-up activities; dissemination of information among investigation participants; public release of information about the investigation; site safety and security; roles of party coordinators, group chairmen, accredited representatives and advisors; identification of the appointed safety board group chairmen who will be allowed at the progress meetings; cockpit voice recorder (CVR) and flight data recorder (FDR); group participation and on-site commander.

Now, the on-site investigation really begins, and specialized groups perform separate investigations. Each of the groups meets daily to review developments in the investigation. Notes are taken, draft reports are discussed, and various theories as to the cause of the crash are explored. The NTSB traditionally relies on manufacturers and air carriers to cooperate and provide them with all relevant information pertinent to the product or operation being investigated.

After several days of investigations, when activities are complete, the IIC will hold a final progress meeting and the command post will be closed. The IIC and the Go Team then return to Headquarters, and a schedule is established for the remainder of the investigation. If no public hearing is required, then all factual information, proposed findings that parties have submitted, petitions for reconsideration and the Board's rulings will be placed in the public docket, which will be made available to the public. If a public hearing is required, then the public docket will be released on the day that the hearing begins.

NTSB public hearings are conducted to obtain more facts. Another purpose for the hearing is to let the public observe the investigation process firsthand. A lot of preparation goes into the planning of a hearing to ensure success. Traditionally, the hearings are held near the accident site. In order to enable the NTSB to determine the probable cause of accidents and improve aviation safety, its investigators are given more legal power than many governmental agencies. For example, NTSB investigators have the right to interrogate witnesses on demand, inspect files, enter facilities and aircraft, and examine the processes and computer data of any party involved in an air crash. Besides these Congressionally authorized powers, the NTSB can obtain subpoenas and court orders for special searches and seizures of any party who may have relevant evidence useful in determining the cause of the air crash.

NTSB regulations specifically state that any witness who is to be "interrogated" has the right to be represented by counsel, which includes the right to be accompanied and advised by an attorney during all aspects of any interrogation in any environment. The NTSB does not generally advise various witnesses or prospective parties of their right to counsel, unless they are actually crewmembers on an accident aircraft. Usually, the air carriers are defendants and are frequently found to be at fault for such crashes. NTSB regulations exclude lawyers for the survivors and insurance companies from participating in the NTSB investigation.

The flight data recorder (FDR), sometimes referred to as the "black box", is actually painted orange so that it can be found more easily in the aircraft wreckage. It records specific flight parameters and is manufactured to withstand enormous impact forces. The data is used in investigations and is also monitored on a regular basis for problems with operations. After the FDR is recovered, it is sent to the FDR Laboratory in Washington DC. The recorder includes the following specific parameters: NTSB or KEYS number, aircraft type, flight number, flight itinerary, number of flights after event, local altimeter setting at the time of the accident/incident, elevation of accident/incident site, location of previous takeoff, runway used, and field elevation, local altimeter setting at time of takeoff, time of departure, coordinated universal time (UTC), time of accident/incident (UTC), accident site conditions that may have caused damage to the recorder (e.g., fire duration, fuel type, etc.). After analysis, the lab sends a preliminary set of plots and tabular data to the IIC. Sometimes, animations of the accident are constructed for better analysis. Typically, though, the FDR is not released until the investigation is complete.

The cockpit voice recorder (CVR) records audio through a microphone that is installed in the cockpit. The CVR is another black box with the same manufacturing parameters as the FDR. Once the CVR is recovered, it is sent to the NTSB Vehicle Divisions Audio Laboratory in Washington DC. Normally, only the CVR specialist, the Directors of the Office of Research and Engineering and the IIC are allowed to listen to the CVR. The laws and policies that govern the procedures regarding CVRs and CVR recordings are generally applicable to any and all audio that is recorded onboard an aircraft. Any audio recording that is recovered from an aircraft following an accident or incident is given the same protection and security of a CVR or CVR recording. Safety Board on-scene staff will secure any device that records audio, found within the cockpit or cabin, carried by a passenger, or installed in the aircraft. Devices that record audio include, but are not limited to, camcorders, video recorders/cameras, digital cameras, handheld tape recorders, personal digital audio recorders, and flight test equipment. Furthermore, any magnetic tape or digital memory chips found in the wreckage could contain recorded audio and are secured by Safety Board staff. Audio from alternate audio devices is not read out or played on scene and the equipment or recording medium is secured by the Safety Board to prevent read-out or damage. In the event that audio from an alternate audio device is recovered, the IIC immediately contacts the Director of the Office of Aviation Safety and the Chief of the Vehicle Recorder Division for guidance.

A transcript of the CVR is a factual record of events. If there is a noise event, it is normally referred to in the transcript as a "sound similar to" something. Typical noise events identified in a transcript include engine sounds, crew seat movement, windshield wiper motors, and aircraft aural warnings. Latched/detent handle movement, such as flaps, slats and gear handles, are sometimes generically identified as a "sound similar to latched/detent handle movement." The IIC is responsible for notifying the surviving flight crew of the opportunity to listen to the recording and review the CVR transcript. The crew review is a courtesy extended to the crew, not a requirement. A CVR specialist supervises the crew's review of the CVR recording and transcript. The crew is not, however, allowed to participate as a member of the CVR group activities. After the crew's review, additions or changes are not to be made to the CVR transcript—changes or comments from the crew are noted in the CVR factual report. Any individual who reviews a CVR recording is bound by Federal CVR nondisclosure laws.

To determine whether pilot fatigue was a factor in the aircraft accident, the investigator asks the following questions: Does the pilot's 72-hour history suggest little sleep or less sleep than usual? Did the accident occur during times of reduced alertness (e.g., 0300 to 0500)? Had the pilot been awake for a long period at the time of the accident? Does the evidence suggest that the accident was a result of inaction or inattention on the part of the pilot? If the answer to any of these questions is yes, then the investigator proceeds with a more detailed methodology. Before concluding that pilot fatigue contributed to an accident, it is important to establish two factors. First, a determination must be made as to whether the pilot was susceptible to fatigue based on sleep lengths, sleep disturbances, circadian factors, and time awake, and/or medical issues. Second, if the pilot was likely experiencing excessive fatigue, then information concerning the pilot's performance, behaviors, and appearance at the time of the accident must be evaluated to determine whether they were consistent with the effects of fatigue. A finding that the pilot was susceptible to the development of a fatigued state in the absence of performance or behaviors consistent with fatigue should not, however, be used to support pilot fatigue as a probable cause or contributing factor in the accident, but may still be an important safety issue to be addressed in the accident report.

Next, the investigator must determine whether the pilot had acute or chronic sleep loss by documenting sleep/wake patterns for at least 72 hours before the accident and learning about the pilot's usual sleep habits. This can be accomplished by asking the pilot the following questions: Describe your typical sleep pattern of when you go to bed, awaken, and how much sleep you get during days off. What time did you fall sleep the night before the accident? What time did you wake up? What was the quality of your sleep? (Repeat for 2 nights before, 3 nights before, etc.) Did you take any naps? When, where, for how long, and why?

Family members, hotel staff or other witnesses who can help complete a description of the pilots' sleep/activity schedule before the accident are interviewed as well. Receipts, cell phone records, work schedules, log books, alarm clock settings, or other records are also used.

The investigator must determine if the pilot's sleep was fragmented (e.g., multiple sleep episodes per 24-hour period) and/or disturbed (e.g., awakenings during sleep due to internal or environmental factors) in days leading to accident. Sleep/wake information is collected and "sleep length" is used to examine the lengths and patterns of sleep episodes for split sleeps or daytime sleep.

The pilot (or pilot's family members) is asked the following questions: Are there factors in your environment (e.g., noise, light, phone calls, etc.) that interfere with your sleep? Was your sleep pattern different or disrupted in the days leading to the accident? The investigator needs to determine if the accident happened during a circadian low point. The primary circadian trough is approximately midnight to 0600, especially 0300 to 0500, while a secondary "afternoon lull" occurs at approximately 1500 to 1700. It must also be determined if the pilot suffered from circadian issues due to recently crossing multiple time zones or to rotating, inverted or variable work/sleep schedules.

The investigator must also determine if sleep disorders or other medical factors (e.g., disease or drug use) were present in the pilot's history. Towards this end, the following questions are asked: Do you have difficulty falling asleep or staying asleep? Have you ever told a doctor about how you sleep? If so, why, when, and what was the result? What drugs/medications do you use regularly, and did you take any in days prior to the accident? Do you have any medical concerns that affect sleep? The investigator will review the pilot's toxicological results for substances that may affect sleep or alertness. If applicable, the pilot will be evaluated using a physician who specializes in sleep medicine. The pilot's medical or pharmacy records are checked, and the wreckage is investigated for any evidence of drugs or medicine. A determination of how long the pilot was awake at the time of the accident is made using interviews or records to estimate wake-up time from most recent significant sleep before the accident. Pilot work records as well as records of previous accidents/incidents (including DMV and/or insurance records) are checked for evidence of prior falling asleep during vehicle operation. A determination is made as to whether representatives of management of labor union parties have indicated complaints of pilot fatigue in the recent past.

The investigator must determine whether the pilot's performance, behaviors, or appearance were consistent with the effects of excessive fatigue, and whether their performance or behaviors contributed to the accident. He or she must also determine whether the pilot's performance was consistent with the effects of fatigue. To accomplish this, the investigator uses available evidence to determine whether the pilot's performance was deteriorating prior to the accident. For example, did the pilot overlook or skip tasks or parts of tasks? Did the pilot focus on one task to the exclusion of more important information? Was there evidence of delayed responses to stimuli or unresponsiveness? Was there evidence of impaired decision-making or an inability to adapt behavior to accommodate new information? Finally, the investigator must determine whether the person's appearance or behaviors before the accident were suggestive of sleepiness/fatigue, as based on witness interviews, pilot report of being tired, audio or video records of the pilot's behavior.

The NTSB hearing for the airliner in the Hudson was held over a three-day period from June 9-11, 2009. The first day included opening statements from the NTSB board member and the IIC.

The rest of the testimony focuses on several topic areas, which vary depending on the nature of the accident. The topic areas during this accident investigation were as follows: Captain and passengers of Flight 1549; pilot training regarding ditchings and forced landings on water; bird detection and mitigation efforts; certification standards regarding ditching and forced landings on water for transport category airplanes; cabin safety; training; procedures and equipment and certification standards for bird ingestion into transport category airplane engines.

In regard to the first topic, Captain and passengers of Flight 1549, Captain Sullenburger was questioned about his decision-making during the accident, his management of the crew, his use of the dual engine failure checklist, and bird strike hazard recognition and avoidance strategies in flight operations. In terms of decision-making, the Captain said he looked for a destination that was long, smooth and reachable. The Hudson River satisfied all three of these requirements. When asked about bird warnings he said "the warnings we get are general in nature and not specific and, therefore, have limited usefulness." He was referring to the Automatic Terminal Information Service (ATIS), which continuously broadcasts takeoff and landing runways, temperature, dew point, ceiling height and other parameters useful to pilots. If there are a large number of birds in the area, the ATIS will contain a warning like "bird advisories in effect". This warning needs to be more specific in order to be helpful – pilots need to know where the birds are, not just that birds are there. One of the passengers also testified about his experience during the flight.

The next topic, pilot training, regarded ditchings and forced landings on water. The aircraft manufacturer testified on the aircraft display indications during a dual engine failure as well as the checklists and manuals that the manufacturer publishes. The airline testified on training and guidance during dual engine failure. Dual engine failure training is given to pilots during their initial training and is covered in annual training as well. A NASA aviation psychologist also testified about how aircrews handle emergencies.

The third topic was bird detection and mitigation efforts. Two members of the USDA testified on the wildlife hazards and their mitigation at airports. Two members of the FAA also testified. The first one testified on trigger events (e.g., a wildlife strike) that require airports to have wildlife hazard assessments (WHA) performed. Once the airport has conducted this assessment, it is required to send the assessment to the FAA. The FAA then makes a determination of the right course of action to take in order to reduce the risk of future collisions between aircrafts and birds.

Many airports hire full-time biologists to control bird populations. Additionally, the airports may have a staff of wildlife technicians available to scare birds away when needed. Anyone in the operational community can call the control tower to dispatch the technicians to the areas of concern and disperse the birds so that they will not cause a hazard. Several years ago, in a European country, an airliner was taxiing out and saw a large flock of birds near the runway. The cockpit recorder even recorded a conversation between the aircrew about the birds. The airliner took off, hit the flock of birds, had a single engine failure and had to return for landing. After a long investigation, the aircrew was faulted for not notifying the control tower so that the birds could be dispersed. This may sound like common sense, but most pilots do not even know that this is an option. Pilots are not trained in methods of wildlife hazard avoidance. There are over 95 airports that, despite having been required to accomplish a WHA due to a strike, have failed to do so.

A second FAA representative testified on the state of bird radar research, reviewing the 8-year history of FAA bird radar research and development. The technology for manufacturing low-cost bird radars was not available when the FAA started this project in 1999, but recently several commercial companies have successfully manufactured and sold these radars. So, in 2007, the FAA changed their strategy from R&D to evaluation of commercial bird radars. The typical cost of an FAA radar that is used to track airliner traffic can be 4 to 6 million dollars. A commercial bird radar, however, can be manufactured for less than one million dollars. The normal use for these bird radars has been to conduct one-time WHAs, but more recently, these commercial companies are trying to adapt the radars to the airport environment. To accomplish the adaptation, the bird radar display is changed so that it shows the runways and surrounding environments, and bird targets are shown with an airport overlay. In 2007, the FAA installed one of these commercial bird radars at Seattle International Airport. The results from the evaluation of this bird radar were discussed as part of the FAA testimony.

The fourth topic of the hearing was certification standards regarding ditching and forced landings on water for transport category airplanes. There were seven witnesses for this testimony. Questions covered the following topics: responsibilities for airframe certification; the history of ditching certification requirements; how the FAA developed ditching standards; how the FAA determines compliance with the standards; assessment of the external damage to the accident airplane (i.e., was it expected or not, and why?); current studies in the area of ditching; criteria used to develop the ditching scenarios (airplane configuration, airspeed, with or without power, etc.); whether the as-designed structure dictated the criteria used or the structure was designed to meet the criteria; and Airbus demonstration of compliance with the ditching requirements for the A320 (e.g., analytical and physical testing, and what airplanes were studied, etc.).

The fifth topic was cabin safety; training; procedures and equipment. The questions asked covered two areas: slide/rafts and emergency training for water landings. In regard to slides/rafts, the questions covered the following areas: history of use on transport category airplanes, certification requirements and standards, assumptions used in A320 certification, comparison of assumptions with other similar airplanes, slide/raft portability, feasibility of use of off-wing slides for flotation, flight attendant training on use of slide/rafts, and an assessment of the cabin damage to the accident airplane (i.e., was it expected or not, and why?). Questions on emergency training for water landings covered the following topics: initial training requirements and drills, recurrent training requirements and drills, planned vs. unplanned events, pre-flight briefings (e.g., on life vests), use of lifelines, and lap children.

The final topic of the hearing was certification standards for bird ingestion into transport category airplane engines. Questions covered the following areas: history of the engine bird ingestion certification requirement, description of the FAA Type Certification process for turbine-powered engines, CFM CFM56-5B4/P compliance with bird ingestion, present and future development of the engine bird ingestion standards, and viability of engine screens. The bird weight requirement for jet engine ingestion had stood at 4 pounds for many years; however, it has recently been increased to 5.5 pounds.[15] Yet the average weight of a Canada goose is 8 to 12 pounds.

Now that the hearings have concluded, the NTSB will start a technical review, which is the final phase of the investigation and includes a final review of all the facts. After this phase, meetings will be held to prepare the final report, which is geared towards a non-technical audience. The report is intended to be used by persons who can make safety changes to the aviation industry so that the same type of accident does not occur in the future. The public, the media and Congress are also part of this audience that is capable of influencing change.

The NTSB determines probable cause, reaches conclusions, and provides safety recommendations. Note that the phrase 'probable cause' does not necessarily refer to the exact cause. This probable cause determination is not based on evidence. The result of a NTSB investigation is a factual report. From this report, the IIC makes a recommendation. The NTSB then reviews the report and votes to adopt, reject or modify the probable cause determination recommended. By law, the NTSB is not required to determine all causes with certainty. After finding the probable cause, NTSB investigations are terminated. The NTSB admittedly does not attempt to allocate responsibility to various parties who may have caused a crash. Unfortunately, the NTSB does not always have the funding and manpower to determine all the causes. When the NTSB approves a report, the probable cause determination, along with a summary of the relevant facts, is published on a quarterly basis in "Briefs of Accident" in the public dockets section of NTSB Headquarters in Washington, D.C. Also worth noting is the fact that NTSB probable cause determinations have been successfully challenged in court and overturned.

The final results of the NTSB investigation are published in

15 Current certification standards for bird ingestion in jet engines are contained in Appendices 2 and 3.

what has been traditionally called The NTSB "Blue Cover" Report (although it is now actually white with blue printing). The NTSB factual report and the NTSB Blue Cover Report are both placed in the public docket at the NTSB Headquarters in Washington, D.C. The only items that are kept out are the trade secrets of parties who have specifically requested that such information be kept confidential and that the NTSB has approved. Unfortunately, the analytical notes, drafts of various reports and other preliminary materials created by NTSB Investigators, NTSB Technical Experts, and party consultants to the NTSB are either destroyed or kept in confidential NTSB files, which are not placed into the public docket.

The final report, though, is available to anyone who wants a copy. All of the exhibits and subgroup reports that were part of the investigations are included in the public docket as well. Representative items that can be found in a public docket from a major accident include weather data, witness statements, cockpit voice recorder transcripts, air traffic control tape transcripts, ground track plots created from FAA radar raw data, engine tear down reports, diagrams, specifications, photographs, computer recreations, transcripts of public hearings, etc. The entire process described above can take 12 to 18 months to complete.

So, now let's relate the process above and discuss the NTSB July 28, 2009 publication of the probable cause of the March 2008 bird strike-related crash in Oklahoma City, Oklahoma of a Cessna 500. The cause was determined to be airplane wing-structure damage that was sustained during impact with one or more large birds (American white pelicans), which resulted in a loss of control of the airplane. The engine was not damaged. On March 4, 2008, at approximately 3:15 p.m. (CST), a Cessna 500, registered to Southwest Orthopedic & Sports Medicine Clinic PC of Oklahoma City, entered a steep descent and crashed about 2 minutes after takeoff from Wiley Post Airport (WPA) in Oklahoma City. Both pilots and the three passengers were killed, and the airplane was destroyed by impact forces and the post-crash fire. The major bird strike safety issues identified by this accident investigation focused on airframe certification standards for bird strikes, inadequate Federal Aviation Administration (FAA) enforcement of wildlife hazard assessment

(WHA) requirements for airports located near wildlife attractants, and lack of published information regarding operational strategies for pilots to minimize bird strike damage to aircraft.

The aircraft performance group looked at the Air Route Surveillance Radar (ARSR), Airport Surveillance Radar (ASR) as well as altitude, position and track data from the transponder. No CVR or FDR data was available, so the NTSB reconstructed the flight from the radar data. They interviewed witnesses and published their testimony. They discovered surveillance video footage from a security camera. They used all the data to create a mathematically modeled six-degrees-of-freedom simulation that estimated the position and orientation of the aircraft throughout the flight.

The structures group created a detailed explanation and display of the aircraft. This group arranged for the wreckage to be shipped to a Texas aircraft salvage location. They laid out the wreckage and described how it was severely fragmented and fire-damaged. They explained in detail the condition of each one of the aircraft surfaces. They could not find any evidence of a bird strike. Then, the structure was examined under a black light, which revealed several areas of fluorescing splatter on the right horizontal stabilizer and the right side of the vertical stabilizer.

The operations group discussed the pilots' ratings and dates of certification, number of flight hours, training and medical history. The aircraft weight and balance and the registered owner were discussed as well. This group also examined FAA oversight and inspection and discovered that the flight was not a legal charter. The maintenance group examined the aircraft logbook records back to 2004 and verified maintenance requirements from the aircraft manufacturer.

So, now that you understand the investigative process, it's time to look into the NTSB as an organization, which leads us to its mission and history. The main purpose of the aviation component of the NTSB is to make air travel safer for all. In 1966, President Lyndon Johnson established the Department of Transportation (DOT), and the Federal Aviation Association (FAA), which had already been in operation and was made a part of it. Then, in 1967, the National Transportation Safety Board was established to conduct independent investigations of all civil aviation accidents in the United

States and major accidents in the other modes of transportation. At that time, the NTSB was funded by the DOT. Then, in 1974, the NTSB severed its ties with the DOT in order to enhance its own status as an independent agency. So, today, the NTSB is not part of the DOT, nor is it organizationally affiliated with any of DOT's agencies, including the FAA. The Safety Board has no regulatory or enforcement powers. And to ensure that Safety Board investigations focus only on improving transportation safety, the Board's analysis of factual information and its determination of probable cause cannot be entered as evidence in a court of law.

The NTSB now has five politically appointed members, including a Chairman, a Vice Chairman, and three others. Each member is nominated by the President and confirmed by the Senate to serve 5-year terms, except for the Chairman and Vice Chairman who serve 2-year terms. Also worth noting is the fact that there can be no more than three members from the same political party. The focus of the NTSB in this book is aviation, however, the NTSB also investigates motor vehicle, boat and railroad crashes.

Reporting to the Board are the Office of General Counsel, the Office of Management, the Equal Employment Opportunity Director, and the Office of Chief Financial Officer. Within the Office of Management are the Office of Research and Engineering; Office of Railroad, Pipeline and Hazardous Materials Safety; Office of Highway Safety; Office of Marine Safety; Office of Aviation Safety; Office of Safety Recommendations and Communications; Office of Administration; and Office of the Academy.

After an investigation, the NTSB submits its recommendation to the Secretary of Transportation, who then has 90 days to respond. The Secretary must also report to Congress once a year on the DOT/FAA's actions in response to the NTSB recommendations. The NTSB has investigated more than 120,000 accidents over a 40-year period and issued more than 12,000 recommendations in all transportation modes. More than 80% of its recommendations have been adopted from various agencies. Since 1973, the NTSB has issued 26 safety recommendations to the FAA and other agencies regarding bird strikes, bird ingestion by jet engines and bird hazard mitigation.[16]

At the NTSB website (www.ntsb.gov), there is an aviation accident database from 1962 to present. Generally, a preliminary report is available within a few days of an aircraft accident. The monthly summary of US civil aviation accidents for the month of July 2009 totaled 156 accidents with 49 fatalities. Eight accidents were from commercial operations and 148 from general aviation. For all of 2008, the total number of accidents was 1,649 with 296 fatalities mostly in general aviation. For January through July 2009, the total number of accidents was 882 with 339 fatalities (a 14% increase over all of 2008)

16 A summary of the NTSB recommendations associated with aircraft bird strike accidents is included in Appendix 4.

Chapter 5:

Bird-Proofing Airliners

Aircraft and helicopters don't need to be completely "bird-proof", however, the empennage (tail), windshield, engines and other aircraft structures must be certified to withstand the impact of a bird. After the impact and resulting damage, the pilot must be able to safely fly and land the aircraft. The airliner in the Hudson was not the only aircraft to lose both engines in flight. The FAA database described in Chapter 3 contains 33 flights over 19 years where bird strikes occurred on both engines of a two-engine aircraft. While the average is 2 per year, recently the numbers have been increasing at an alarming rate. In just the last two years alone, there were 7 two-engine bird strikes – 3 in 2007 and 4 in 2008.

In the final report for the Oklahoma City accident that occurred in 2008,[17] the NTSB recommended that the FAA improve aircraft standards for withstanding bird strikes. These recommendations are included as the last three entries in Appendix 4. The current standards were developed 40 years ago, and several factors have influenced this recommendation to improve upon the original standards. First, as mentioned before, environmental protection laws from the 1960s and 70s have caused the populations of large bird

17 Please see Chapter 4.

species to dramatically increase. Second, the number of passengers traveling has dramatically increased and is forecast to continue to grow from a current annual total of 600 million (nearly 2 million/ day) to 1 billion in 2020. All of this means that there are more birds and more planes sharing the same amount of sky, so it's just a matter of time before we have another severe two-engine failure due to multiple birds striking an aircraft.

Another recent accident where both engines were lost due to ingestion of birds occurred to a Boeing 737, only two months before the Flight 1549 accident, on November 10, 2008 at Rome Ciampino Airport in Italy. The airplane experienced multiple bird strikes to both engines while on approach, and an engine flamed out as a result. The crew initiated a go-around, but at about the same time, the second engine flamed out as well, so the aircraft landed and veered off the runway during rollout. Extensive damage occurred to the left wing, the left main landing gear and the belly of the fuselage. Passengers reported that they noticed a loud bang followed by violent vibrations of the airplane and, just moments before touch down, a burning smell in the cabin. Blood and feathers were visible on some passenger windows along the fuselage. One passenger said that he saw flames shooting out of one engine. Passengers' descriptions of the actual touchdown ranged from "firm" to "slamming the aircraft down".

Military aircraft have also been involved in serious bird strike accidents. The most famous one that I can recall was the loss of an E-3 AWACS (Airborne Warning and Control System), which is a Boeing 707 aircraft with a very powerful airborne surveillance radar. On takeoff at Elmondorf AFB, AK, the aircraft struck a flock of Canada geese. Both engines on the left side of the aircraft suffered multiple bird strikes and lost power. The aircraft rolled left and crashed into a hill. All 24 crewmembers were killed. About a year after that, another NATO E-3 AWACS was rolling down the runway at a Greek airbase when it encountered a flock of birds. The pilots elected to accomplish a high-speed abort, and the aircraft ran off the runway and ended up in the water. The fuselage broke apart but all crewmembers survived.

When a jet strikes large flocking birds at very high speeds, severe damage can result. The parameters that affect the force of the impact include the bird weight, density, speed and angle of impact. The force of impact can be estimated using a very simple calculation. From physics, we have this equation: kinetic energy equals force times distance. Kinetic energy also equals one half of the mass times the velocity squared. So basically, force is just kinetic energy divided by distance. In other words, the greater the mass of the bird, the higher the impact force. Since the impact force increases by the square of the airspeed, the impact speed is the parameter that causes the greatest damage. The higher the speed, the greater the force. For a jet aircraft at 200 knots, a 4-pound bird will impact with a force of 4.8 tons. If the speed is increased to the maximum speed allowed below 10000 feet (i.e., 250 knots), then the force increases to 7.5 tons.

It is often impossible for a pilot to notice a bird before the aircraft collides. The aircraft is traveling at a high speed, and the bird is very small and almost impossible to see. Specific bird parameters that affect the extent of the damage include bird weight, density and impact angle.

Federal Aviation Regulations (FARs) contain certification standards that require both the airframe and the engine manufacturers to demonstrate acceptable levels of performance during a bird strike. Appendix 2 contains initial requirements for foreign object ingestion in a jet engine, and Appendix 3 has a complete description of a recent amendment to the 40-year-old standards for bird ingestion in Appendix 2. The FARs are reviewed on a periodic basis, and changes are sometimes recommended after the NTSB completes their investigations. There are two agencies that create certification standards – the FAA and the European JAA (Joint Aviation Authority). They try to work together to ensure that the standards apply worldwide.

Normally, one company manufactures the aircraft and another produces the engine. These companies have made tremendous improvements in material and structural designs. These improvements, in turn, allow an aircraft or engine to take a severe impact and continue to safely fly and land. If the bird strike causes the loss of a single engine, the aircraft are certified to continue flight on one engine. Pilots receive initial and recurrent training to maneuver an aircraft on one engine, so they are always prepared for single-engine flight.

When a bird is ingested by a jet engine, FAA regulations require that the engine can be safely shut down. The engine can come apart, but this type of failure must be contained within the engine cowling. No parts of the engine can be allowed to escape and impact other parts of the aircraft. Qualification to meet these standards includes both simulation and testing by means of propelling dead birds with specified weights and quantities into working engines.

The following are portions of FAR Part 25 that describe the structural requirements for transport category aircraft to withstand bird strikes:

The Airframe

> Certain parts of the airframe require certification by federal regulations. Currently, aircraft manufacturers are using composite material in their designs to lighten the weight and, as a result, decrease fuel consumption. Windshield strength has improved by using new materials and heating the material to increase flexibility. The USAF is developing injection-molded frameless canopies for fighters. This will make the canopy lighter and improve maintainability.

The Empennage

> The empennage structure must be designed to assure capability of continued safe flight and landing of the airplane after impact with an 8-pound bird when the velocity of the airplane (relative to the bird along the airplane's flight path) is equal to V_c at sea level, selected under Part 25.335(a). Compliance with this section by provision of redundant structure and protected location of control system elements or protective devices, such as splitter plates or energy-absorbing material, is acceptable. Where analysis, tests, or both show compliance, use of data on airplanes having similar structural design is acceptable.

The Windshields and Windows

Internal panes must be made of non-splintering material. Windshield panes directly in front of the pilots in the normal conduct of their duties, and the supporting structures for these panes, must withstand, without penetration, the impact of a 4-pound bird when the velocity of the airplane (relative to the bird along the airplane's flight path) is equal to the value of V_c, at sea level.

The Structure of the Rest of the Aircraft

An evaluation of strength, detailed design, and fabrication must show that catastrophic failure due to fatigue, corrosion, or accidental damage will be avoided throughout the operational life of the airplane. This evaluation must be conducted for each part of the structure that could contribute to a catastrophic failure (e.g., wing, empennage, control surfaces and their systems, fuselage, engine mounting, landing gear, and their related primary attachments). The airplane must be capable of successfully completing a flight during which likely structural damage may occur as a result of (a) impact with a 4-pound bird when the velocity of the airplane relative to the bird along the airplane's flight path is equal to V_c at sea level or 0.85 V_c at 8,000 feet, whichever is more critical; (b) uncontained fan blade impact; (c) uncontained engine failure; or (d) uncontained high-energy rotating machinery failure.

For a Normal, Utility, Acrobatic and Commuter Aircraft, FAR Part 23 applies. Windshield panes directly in front of the pilots in the normal conduct of their duties, and the supporting structures for these panes, must withstand, without penetration, the impact of a 2-pound bird when the velocity of the airplane (relative to the bird along the airplane's flight path) is equal to the airplane's maximum approach flap speed. The windshield panels in front of the pilots must be arranged so that, assuming the loss of vision through any one panel, one or more panels remain available for use by a pilot seated at a pilot station to permit continued safe flight and landing. And finally, helicopter bird strike requirements are listed in FAR Part 29. A rotorcraft must be designed to ensure capability of continued safe flight and landing (for Category A) or safe landing (for Category B) after impact with a 2.2-pound bird when the velocity of the rotorcraft (relative to the bird along the flight path of the rotorcraft) is equal to V_{NE} or V_H (whichever is less) at altitudes up to 8000 feet. Compliance must be shown by tests or by analysis based on tests carried out on sufficiently representative structures of a similar design.

The recent Oklahoma City accident described in Chapter 4 involved the destruction of an aircraft due to a large American pelican (15 pounds, 10-foot wingspan) hitting the wing of a small corporate jet and causing wing structure damage. There was no damage to the engine. As a result of this fatal accident, the NTSB made the last three recommendations in Appendix 4. The first of them reads: "Revise the bird-strike certification requirements for 14 Code of Federal Regulations Part 25 airplanes so that protection from in-flight impact with birds is consistent across all airframe structures. Consider the most current military and civilian bird-strike database information and trends in bird populations in drafting this revision."

The reason for this recommendation is that the current design requirements for bird strikes to the airframe are inconsistent. The inconsistency between the regulations can be seen by examining 14 CFR (Code of Federal Regulations Part 14) Part 25-571, which is the same as European Aviation Safety Agency (EASA) CS-25.631; this requirement for the design and construction of transport category airplanes reads as follows: "The *empennage* structure must be designed to assure capability of continued safe flight and landing of the airplane after impact with an 8-pound bird when the velocity of the airplane (relative to the bird along the airplane's flight path) is equal to V_c at sea level." An additional requirement for a bird strike to the *rest of the aircraft structure* is for the aircraft to be capable of continued safe flight and landing after hitting a 4-lb. bird at the more critical speed of V_c (cruise speed) at mean sea level or 85% of V_c at 8000 feet altitude. To confuse matters even more, there is a third requirement for the *windshield*. Both EASA CS-25 and 14 CFR Part 25 require that windshield integrity after single bird impact result in the inner ply being non-splintered and the panes directly in front of the pilots must withstand, without penetration, a 4-pound bird at cruise speed at mean sea level and pitot tubes (an instrument that measures airspeed) must be far enough apart to preclude damage from a single bird impact. Under EASA CS-23.775 and 14 CFR Part 23.775, smaller aircraft are required only to have limited windshield integrity – a demonstrated single bird impact resistance of up to 2 pounds at maximum approach flap speed and at least one pane with sufficient forward vision remaining to allow continued safe flight. To resolve these inconsistencies, the NTSB is recommending that the FAA examine these inconsistent numbers and come up with a single value for the bird weight based on data in the FAA bird strike database.

The other applicable NTSB recommendation reads: "Require aircraft manufacturers to develop aircraft-specific guidance information that will assist pilots in devising precautionary aircraft operational strategies for minimizing the severity of aircraft damage sustained during a bird strike, should one occur, when operating in areas of known bird activity. This guidance information can include, but is not limited to, airspeed charts that depict minimum safe airspeeds for various aircraft gross weights, flap configurations, and power settings; and maximum airspeeds, defined as a function of bird masses, that are based on the aircraft's demonstrated bird-strike energy." This recommendation is very easy to interpret as the needed items are specifically listed.

Because of the airliner ditching in the Hudson, Congress and the NTSB have highlighted, but not required, stricter regulations for bird strike damage certification. Some aircraft manufacturers are trying to stay ahead of the regulation changes that may occur when the final report is issued. According to John Croft of *Flight Global*, Boeing recently announced that it would increase the bird strike tolerance of its newly remodeled next-generation Boeing 737. A new windshield will be designed to keep loose glass from entering the flight deck in the event of a bird strike. For the windshield, manufacturers must either show by analysis or tests that the likelihood of the glass fragmenting as a result of the impact and hitting the pilots during the event is "of low order" or they must "have a means to minimize the danger to the pilots from flying windshield fragments due to bird impact," according to the regulations.

The redesigned laminated glass windshield will be slightly smaller and will include an inboard antispall liner to prevent broken glass from entering the flight deck during a bird strike event. Adding an antispall liner to the windshields for Next-Generation 737 airplanes enables Boeing to keep the structural airframe design while incorporating newer technology. Certification of the new design for the windshields is scheduled for the second quarter of 2010.

The Engine

In 1970, the certification procedures for foreign object ingestion in jet engines were published in an advisory circular (AC). An explanation for the purpose of an advisory circular is included in the FAA advisory circular checklist dated June 15, 2000. "The FAA issues advisory circulars to inform the aviation public in a systematic way of non-regulatory material. Unless incorporated into a regulation by reference, the contents of an advisory circular are not binding on the public. Advisory circulars are issued in a numbered-subject system corresponding to the subject areas of the Federal Aviation Regulations (Title 14, Code of Federal Regulations, Chapter I, FAA). An AC is issued to provide guidance and information in a designated subject area or to show a method acceptable to the Administrator for complying with a related Federal Aviation Regulation."

Advisory circular AC 33-1B dated April 22, 1970 in Appendix 2 is titled "Turbine Engine Foreign Object Ingestion and Rotor Blade Containment Type Certification Procedures." These procedures include "desirable engine design features like blades which effectively mince birds upon contact." Other items classified as foreign objects in the regulation include the following: a cleaning cloth, a mechanic's hand tool, a small-size aircraft steel bolt and nut typical of aircraft inlet hardware, a piece of aircraft tire tread and compressor and turbine rotor blades. Other foreign objects include water in the form of rain and snow, gravel of mixed sizes up to one-fourth inch, sand of mixed sizes, ice of typical sizes and forms representative of inlet duct and lip formations, engine front frame and guide vane deposits, in quantities likely to be ingested during a flight, hail stones of approximately 0.8 to 0.9 specific gravity and of one- and two-inch diameter, birds in weight categories including small birds of two to four ounces (e.g., starlings) and medium birds of one to two pounds (e.g., common gulls, small ducks, and pigeons). The regulation classifies birds with weights of 4 lbs. and over (e.g., geese, buzzards, largest gulls and ducks) as foreign objects that could damage a jet engine.

Most of the time, large birds were encountered in flocks and rarely as single birds. Ingestion testing the large bird is aimed primarily at substantiating direct impact effects. Ingestion testing of the smaller birds is aimed at substantiating the effects of various bird masses. Both inlet opening width and overall area have a bearing on the probability of ingestion of given size birds, and these factors, along with bird flocking density, were considered in selecting bird sizes and quantities.

The AC also formulated the initial concepts for testing to comply with the standards. The AC states that "bird ingestion tests should use freshly killed birds and states that gun injection is preferable as a simulated strike." The AC also states that other "acceptable techniques have been used which utilize previously frozen birds and injection means other than guns. If previously frozen birds are used, they should be completely thawed for the tests, and have normal moisture content. If frozen for appreciable periods, moisture content may be reduced below normal levels. Use of synthetic "birds" has been proposed and will be acceptable if the results of ingestion can be shown to be equivalent to ingesting actual birds. For testing impact effects, include all frontal areas considered to be critical, and appropriate bird velocities should be attained at the inlet. Other ingestion effects, such as compressor stall or blowout, may be sufficiently severe at somewhat lower bird velocities.

Other initial requirements from the AC are as follows: "When small birds are used, they should be ingested at typical takeoff flight speeds and engine output levels. One small bird should be ingested for each 50 square inches of inlet area (or fraction thereof) if it can enter the inlet and reach the engine face. The maximum number of small birds ingested as a group need not exceed 16. Small bird testing may be omitted for large engines when it is agreed that medium birds will pass into the engine blading passages and result in a test of at least equal severity. When medium-sized birds are used, ingest at a typical initial climb speed with takeoff engine output. When medium birds enter the inlet and reach the engine intake section, ingest one bird for each 300 square inches (or fraction thereof) of intake area up to 3,000 square inches, with additional medium birds at 1/3 of this rate for larger engines. Small and medium birds should be ingested in random sequence, dispersed over the inlet area, to simulate an encounter with a flock."

Current standards for both multiple- and single-bird engine ingestions into a single fixed-wing aircraft engine exist in equivalent form in Part 14 Code of Federal Regulations (CFR), Parts 33-77, and in the European Aviation Safety Agency (EASA) Airworthiness Code CS-E 800 'Bird Strike and Ingestion'. The basic requirements for engine ingestion were revised in 2000 to take into account both evidence of an increase in the size of birds impacting aircraft and issues raised by the development of very large inlet, and high-bypass-ratio, engines. The requirements are summarized by SKYbrary, an initiative of EUROCONTROL, International Civil Aviation Organization (ICAO) and the Flight Safety Foundation, as follows:

a) At a typical initial climb speed and takeoff thrust, ingestion of a single bird of maximum weight between 4 pounds and 8 pounds, dependent upon engine inlet area, shall not cause an engine to catch fire, suffer uncontained failure or become impossible to shut down and shall enable at least 50% thrust to be obtained for at least 14 minutes after ingestion. These requirements are to be met with no thrust lever movement on an affected engine until at least 15 seconds have elapsed post impact.

b) At a typical initial climb speed and takeoff thrust, ingestion of a single bird of maximum weight of 3 pounds shall not cause a sustained thrust or power loss of more than 25%, shall not require engine shut down within 5 minutes and shall not result in hazardous engine condition.

c) At a typical initial climb speed and takeoff thrust, simultaneous ingestion of up to 7 medium-sized birds of various sizes between a weight of 0.77 pounds and 2.5 pounds, with the number and size depending upon the engine inlet area, shall not cause the engine to suddenly and completely fail, and it shall continue to deliver usable but slowly decreasing minimum thrust over a period of 20 minutes after ingestion. (Engines with inlet sizes of less than 300 square inches only have to meet the standard for a single bird of this weight.)

d) At a typical initial climb speed and takeoff thrust, simultaneous ingestion of up to 16 small-sized birds of weight 0.187 pounds, with the number dependent upon the engine inlet area, shall not cause the engine to suddenly and completely fail, and it shall continue to deliver usable but slowly decreasing minimum thrust over a period of 20 minutes after ingestion.

During the Congressional testimony for Flight 1549, certification for the jet engines on the Airbus A320 was discussed. Many years ago, the requirement for bird ingestion was for a 1.5-pound bird directed at the core of the engine, which is its center. In an airliner jet engine, most of the air (85%) bypasses the core to improve jet engine efficiency. In a fighter aircraft, most of the air goes through the core of the engine. Most of the time, a bird gets caught in the bypass area of the engine and the engine continues to run because the bird did not penetrate the core. The reason for the requirement for the bird being aimed at the core is to simulate a worst-case scenario. A bird going through the core will normally damage the critical engine components, and the engine will seize, come apart or shut down. Today, the FAA requirement for testing the Airbus A320 engine for a single-bird ingestion is four pounds. After ingestion, the engine is not required to continue to run. The requirement is to safely shut down without causing hazard or fire. For a flock of birds, the requirement is different. This engine is required to ingest seven 1.5-pound birds and must continue to run for five minutes at its takeoff power setting. After five minutes, the engine is allowed to lose 25% of its thrust. Out of the 650 bird species that live in the United States, 36 have an average body mass greater than 4 pounds. It is obvious from the successful ditching of Flight 1549 that the aircraft and engines performed as expected. When you factor in the average weight of a Canada goose as 8 to 12 pounds, you realize that the aircraft and engines performed in excess of the requirements. After the engines ingested the birds, they did not come apart, and no parts damaged the aircraft.

The NTSB is investigating whether certification standards for jet engines that sustain bird strikes need to be changed. For engines that sustain strikes from what are termed "large flocking birds", current standards require that a safe engine shutdown without fire or the expelling of debris is necessary, according to testimony. The FAA standards do not require that the affected engines maintain thrust. The NTSB board, which will issue a final report on the US Airways Flight 1549 crash early in 2010, could consider recommending to the FAA that engine certification standards be changed. Changes to increase the structural integrity of the jet engine will significantly increase weight, which is an undesirable quality since fuel consumption is such a critical concern these days.

I have talked to several people who ask "Why can't you just place a screen over the engine to protect it from birds?" Well, this would be impractical for several reasons. First of all, the screen would cause an airflow problem with the engine. Turbulent air could stall the engine and would certainly decrease fuel efficiency. Another issue is flight in icing conditions. Moisture would build up on the screen and eventually break off, be ingested into the engine, and possibly cause the same problem that you are trying to avoid.

Recently, ALPA led efforts to improve jet engine bird ingestion certification standards. In 1976, the NTSB made a recommendation that called on the FAA to revise the regulations "to increase the maximum number of birds in the various size categories required to be ingested into turbine engines with large inlets." The FAA agreed that the engine certification regulations should be modified to expand the bird ingestion testing requirements for large, high-bypass-ratio engines and began to improve the standards. Although these rulemaking efforts represent progress in engine certification, the data used to develop the current standard is now over 40 years old. During the intervening years, aggressive conservation programs have resulted in increased populations of larger birds. Continued avian research to determine today's bird strike risk is needed to improve engine certification standards.

Now that we have discussed the regulations and bird strike requirements, it is time to determine how to validate requirements. To verify that an aircraft can meet the requirements, analysis and simulation can be used. Bird strike simulation is very complex and requires the use of computational modeling, nonlinear dynamics, flow structure interaction and failure modeling. In order to perform the simulation, it is necessary to create a model for the bird, the blades of the jet engine, and the rest of the aircraft structures.

One bird strike simulation method is a computer-based application called LS-DYNA. Equations are solved by using mechanical laws of conservation of mass and conservation of momentum to obtain the velocity, density, and pressure of the fluid (i.e., air) for a specific position and time. Impact problems are normally broken down into two categories – hard-body (no deformation) and soft-body (deformation) impact. A bird impact is obviously a soft-body problem because it will deform on impact. The material properties used to model a bird include fluid and air, using a water and air mixture. For the bird model, a material, shape, viscosity, and mass must be chosen. To model the aircraft structure, a material, shape, strength, elasticity and density must be chosen. Outputs from the simulation include pressure vs. time and force vs. time. Once the plots are obtained, an analysis takes place to determine if the aircraft component will fail. If a failure is predicted to occur, then the aircraft manufacturer may need to change the design or material used to construct that particular aircraft structure. LS-DYNA is widely used by the aviation industry to simulate bird strikes, to verify that engine blades will remain contained, and to verify that aircraft components will not suffer a structural failure. Validated techniques have been developed that enable the accurate prediction of the structural deformations and the movement of the bird over the structure.

Another way to validate high-speed bird strike requirements is live testing, but this is much more expensive than simulation. A device called a "chicken gun" is normally used to conduct this sort of test. Other names for this device are the "chicken cannon", "turkey gun" or "rooster booster". The device was first used in the United Kingdom in the 1950s and has been used in the United States since 1972. Most chicken guns are owned by the aircraft manufacturers. The National Research Council (NRC) in Canada owns a chicken gun and is always open for business. Since 1968, NRC has fired its 10-inch-bore chicken gun more than 3,500 times, consuming more than 3.5 tons of chickens in the process. NRC staff once completed 55 shots in 4.5 days. In addition to the 10-inch gun, NRC also operates 3.5-inch and 5-inch cannons for "engine ingestion" tests. Recently, the NRC chicken gun was used to certify the CVR and the FDR for impact.

After a break up of the Space Shuttle in 2004, USAF Arnold Engineering Development Center engineers used a chicken gun to help the Shuttle return to flight. NASA used the facility to shoot foam projectiles at various velocities and angles at the struts that connect the solid rocket booster and external fuel tank, core panels and cover material for the range safety system antennas. High-pressure helium gas was used to launch foam projectiles at speeds between 102 and 1,537 mph down an 86-foot-long rectangular barrel. High-speed video cameras operating at up to 20,000 frames per second were used to measure velocity. Strain gauges and accelerometers were used to determine stress levels.

The chicken gun is a large-diameter (about 10 inches) long-barrel (40-75 feet) cannon filled with compressed air (up to 200 psi). You just load the gun with a bird, wait a few minutes, and fire it. The set-up normally takes about 20 minutes. The chicken gun rests on a heavy frame that can be aimed by adjusting it up and down and side to side. Both the FAA and the USAF use the chicken gun to test the strength of windshields, nose cones, wings and jet engines. This device is named after its projectile, which is a dead chicken of the type that can be obtained in a supermarket for cooking. The method used for testing is to fix an aircraft or portion of an aircraft on a test stand, and then use the chicken gun to fire a chicken into the windshield, engine or other aircraft structure. Slow-motion cameras are normally used to videotape the impact. In addition to the engine, the birds are fired at the canopies, windshields, wings and empennages.

When the chicken gun is used to test a jet engine, a large closed room inside of a building is normally used to enclose the jet engine. Jet engines are normally tested to obtain FAA certification. The bird is normally fired at a speed of approximately 180 mph. Because USAF aircraft fly at high speeds at low altitudes, they normally fire birds at fighter aircraft canopies in excess of 400 mph. Other components of a fighter aircraft are tested at speeds of up to Mach 1 (supersonic). Some companies take the process one step farther and purchase birds that still have their feathers attached. Most companies buy the birds frozen in the supermarket and thaw them in the microwave just before the test. The temperature of the bird is taken before firing to prove to the customer that it was not frozen. The engine is operated for twenty minutes after the impact, and if it disintegrates, it fails the test.

There is an old story about the British being interested in using a chicken gun to test a windshield for their new high-speed trains. Arrangements were made, and they obtained a chicken gun from NASA. When the gun was fired, the engineers stood shocked as the chicken hurtled out of the barrel, through the shatterproof shield, smashing it to smithereens, through the control console, snapping the engineer's backrest in two and embedding itself in the back wall of the cab. Horrified Britons sent NASA the disastrous results of the experiment, along with the designs of the windshield, and begged the US scientists for suggestions. NASA's response was short: "Thaw the chicken."

Chapter 6:

Organizations Working on the Bird Strike Problem

NTSB and FAA

Since 1973, the National Transportation Safety Board (NTSB) has been concerned about aircraft bird strikes. Whenever an aircraft accident occurs, the NTSB is the organization that is called in to perform a very thorough investigation. If the accident is very serious, then the NTSB publishes a safety recommendation. A complete list of all NTSB recommendations concerning aircraft bird strike accidents from 1973 to present is contained in Appendix 4.

On March 22, 1995, the FAA published an order on how to handle a NTSB safety recommendation. Both organizations share a common objective – to promote aviation safety and prevent aircraft accidents. The FAA is usually a participant in NTSB investigations. The order states that the FAA will consider current NTSB recommendations and review the status of past recommendations. The FAA's office of Accident Investigation (AAI) serves as FAA's focal point for receiving, processing, managing, and tracking NTSB safety recommendations and as the liaison between FAA and the NTSB on all NTSB safety recommendation issues. The AAI office also prepares initial FAA responses to NTSB safety recommendations for the Administrator's signature no later than 70 days after receipt of the NTSB safety recommendation. Basically, the order makes it mandatory for the FAA to respond to NTSB safety recommendations, but it does not, however, require that the FAA take action.

The first NTSB safety recommendation for an aircraft bird strike appeared in 1973, and the next recommendation did not occur until 23 years later. The environmental laws were passed in the 1960s and 70s, and as a result, the bird population began to increase, but it took a while for it to increase to the extent that it has today. Add to that the dramatic increase in air travel, and you have the perfect storm. Predictably, many severe aircraft bird strike accidents occurred in the 1990s. There were two accidents that broke the camel's back. The first was an accident on June 3, 1995, when an Air France Concorde, at about 10 feet AGL, while landing at John F. Kennedy International Airport (NY), ingested 1 or 2 Canada geese into the #3 engine, which then suffered an uncontained failure – shrapnel from the #3 engine destroyed the #4 engine and cut several hydraulic lines and control cables. The pilot was able to land the plane safely, but the runway was closed for several hours. Damage to the Concorde was estimated at over $7 million. The French Aviation Authority sued the Port Authority of New York and New Jersey and eventually settled out of court for $5.3 million. The second accident was more dangerous – in 1995, an E-3 crashed from bird ingestion into two of its engines, resulting in total loss of life.

After the final reports for these accidents were released in 1996, several major changes were made as an attempt to fix the aircraft bird strike problem. The NTSB published five aircraft bird strike-related safety recommendations. The first was A-96-38, advising pilots to delay takeoff whenever a bird hazard exists. Then, in A-96-39, the NTSB asked the FAA to develop a set of scare tactic procedures to disperse birds from near runways. In A-96-40, the NTSB asked the FAA to annually brief air traffic controllers on the dissemination of bird hazard information to pilots. In A-96-41, the NTSB asked the FAA to provide bird-specific hazard information on the Automatic Terminal Information System (ATIS). In A-96-42, the NTSB asked the FAA to produce a form for pilots to report bird strikes. None of these recommendations constituted a permanent fix; they were simply ways to reduce risk. Concerning the dissemination of information to pilots, I can say, as an airline pilot myself, that an ATIS warning like "bird activity in the vicinity of the airport" is not helpful. First of all, birds are basically everywhere, and secondly, what am I supposed to do with this warning if I can't see the birds? The most I can get out of this warning is that I have a greater chance than usual of hitting something but still can't do anything about it.

Over the next few years, several more severe bird strike accidents occurred. On January 7, 1997, an MD-80 aircraft struck over 400 blackbirds just after takeoff from Dallas-Fort Worth International Airport (TX). Almost every part of the plane was hit. The pilot declared an emergency and returned to land without event. Substantial damage was found on various parts of the aircraft, and the #1 engine had to be replaced. The runway was closed for 1 hour. The birds had been attracted to an un-harvested wheat field on the airport. Almost exactly one year later, on January 9, 1998, while climbing through 3000 feet, following takeoff from Houston Intercontinental Airport (TX), a Boeing-727 struck a flock of snow geese with 3-5 birds ingested into 1 engine. The engine lost all power and was destroyed. The radome was torn from the aircraft and the leading edges of both wings were damaged. The pitot tube for first officer was torn off. Intense vibration was experienced in the airframe, and the noise level in cockpit increased to the point that communication among crewmembers became difficult. An emergency was declared. The flight returned safely to Houston with major damage to the aircraft.

After several serious aircraft bird strike accidents in 1999, the NTSB recommended that the FAA, in coordination with the US Department of Agriculture, conduct research to "determine the effectiveness and limitations of existing and potential bird hazard reduction technologies." This recommendation kicked off an FAA research and development project to find a technical solution to the aircraft bird strike problem. Radar research for the FAA project started at MIT Lincoln Laboratory in 2000. In my opinion, the FAA has been nibbling away at the aircraft bird strike problem by implementing the NTSB recommendations listed in Appendix 4. Ten years later, there is still no permanent technical solution despite the fact the technology (i.e., radar) exists to solve it.

Air Line Pilots Association (ALPA)

ALPA is a pilots' union that represents over 65,000 airline pilots and has been directly involved with the resolution of aircraft bird strike issues. In 2000, ALPA recommended that Congress, the FAA, and other US agencies take the following actions to reduce wildlife hazards to aviation:

- Congress should authorize at least $450,000 per year for several years to reduce wildlife hazards to aviation and manage the FAA's wildlife strike database.
- Congress should authorize at least $600,000 per year for several years for the federal government to conduct wildlife surveys at airports.
- All pertinent members of the aviation industry and the US government should be educated on this problem. Flight crew members should receive formal education on wildlife hazards during their annual recurrent training, along with guidance on how to reduce these hazards. This training should be similar to FAA-mandated training on windshear, bomb threats, de-icing, and other aviation hazards. Airport certification inspectors should receive formal training on plans for reducing the amount of wildlife so that airport inspections require evidence of an effective plan. Air traffic controllers should review, in their periodic training, FAA Order 7110.65, paragraph 2-1-22.
- Immediately, reporting of wildlife strikes or hazards should be required, as is mandatory in most western European countries. Any party – pilot, mechanic, airport operations personnel, or air traffic controller – having evidence of wildlife hazards should report it. An FAA wildlife strike database and a reporting system already exists and can handle a greater rate of reporting. Although every air carrier has a safety department that collects similar data, only two of them are reporting their strikes to the FAA.

- All airport managements should immediately survey their airports to identify wildlife attractants on airport grounds. Federal funds should be made available to help minimize these attractants.
- The FAA should issue an advisory circular on reducing wildlife hazards on airports, similar to the wildlife hazard documents developed by Transport Canada.
- Congress should appropriate funds for wildlife hazards research. This research and its funding should be administered by the USDA.
- The FAA Technical Center should spend the $800,000 that Congress authorized, instead of the $200,000 that the FAA has elected to spend, for wildlife hazard research and mitigation.
- The FAA, the USDA, the Environmental Protection Agency, the US Army Corps of Engineers, and the Fish and Wildlife Service should immediately enter into an intra-government agreement to expeditiously resolve public safety matters when federal rules or laws conflict, such as those governing airport wetlands, which serve as wildlife attractants. The agreement should also designate airports and the areas around those airports as "wildlife incompatible" for the protection of both the traveling public and animals.
- Engine certification standards should reflect the current and foreseeable threat. Some methods of increasing engine robustness in the future, as aviation and wildlife populations grow, should be included in the Notice of Proposed Rulemaking (NPRM) on engine certification for bird ingestion that the FAA currently is considering.
- The FAA should abandon its operational evaluation that permits and encourages airspeeds greater than 250 KIAS below 10000 feet in areas of known bird activity. The severe damage to the Delta B-727 that departed Houston under this program illustrates the dangers of high-speed bird strikes. Until aircraft and engines are strengthened or other mitigation actions are taken, high-speed flight at low altitude should be avoided.

- The FAA should use all technology currently available – for example, Nexrad radar and approach control radars – to warn pilots of imminent hazards. Air traffic controllers should be educated about this threat to public safety and should be required to comply with FAA Order 7110.65 paragraph 2-1-22 and issue timely warnings to pilots, just as they issue windshear alerts, braking action reports, and other safety advisories.

The Associated Press (AP) recently reported "It's been 10 years since the National Transportation Safety Board recommended the Federal Aviation Administration develop a radar system that enables airline pilots to avoid birds. The FAA is testing experimental systems at some airports, but agency officials caution the technology is unproven and still needs years of refinements." When interviewed by AP, Rory Kay, a Boeing 767 captain and Safety Chairman of ALPA, told reporters "that is not a satisfactory timeframe," and ALPA included the bird problem among the union's top safety priorities for 2009. At a press conference announcing these priorities, Capt. Kay pointed made the following statement:

> ...when airports have tried to develop high-speed departure procedures, ALPA has been there to point out the dangers associated with bird strikes. As a result of this participation, the FAA's speed limits at low altitude (below 10000 feet) have been maintained. APLA safety representatives were involved in the amendment of the certification standards for aircraft engines, which resulted in more stringent requirements for engine design. ALPA has also been directly involved in the development of wildlife mitigation plans at commercial airports. Meanwhile, the FAA is evaluating the use of low-cost radar to detect birds within 3 to 5 miles of an airport and develop an airport bird strike advisory system that could relay warnings directly to airline cockpits. ALPA will continue to monitor this research and push for widespread use if the system proves worthy.

Bird Strike Committee USA

Within the United States, there was no single forum where information or concerns dealing with this problem could be addressed. Bird Strike Committee USA was formed in 1991 for several reasons: to facilitate the exchange of information; to promote the collection and analysis of accurate wildlife strike data; to promote the development of new technologies for reducing wildlife hazards: to promote professionalism in airports' wildlife management programs through training and advocacy of high standards of conduct for airport biologists and bird patrol personnel; and to act as a liaison to similar organizations in other countries. Bird Strike Committee USA is a volunteer organization directed by a 10- to 15-person steering committee consisting of 2-3 members each from the FAA, USDA, Department of Defense, aviation industry/airlines, and airports. Typically, a Bird Strike Committee USA meeting takes place over the course of 3 ½ days and consists of four parts: Part 1 is practical classroom and field training sessions on wildlife control at airports, covering both civil and military aviation; Part 2 consists of the presentation of technical papers and posters; Part 3 involves exhibits and demonstrations with vendors; and Part 4 is a field trip that generally covers the host airport or a wildlife management area to observe management programs and habitat issues related to wildlife and aviation safety.

A Bird Strike Committee meeting covers the following subjects:

- Wildlife strike reporting/statistics in relation to safety management systems
- Bird management and control techniques
- Research on new technologies to reduce wildlife hazards
- Training in wildlife management on airports
- Military concerns of wildlife hazards
- Aircraft engines/components performance and standards related to wildlife hazards
- Policies/standards for airports and aircraft operations related to wildlife hazards
- Land use and environmental issues concerning airports

- Avian migration, behavior and sensory capabilities related to aviation
- Remote sensing/modeling to detect and predict bird numbers and movements

Participation in the annual meetings is open to any person interested in reducing wildlife hazards to aviation or in wildlife and environmental management at airports. As examples, recent meetings have been attended by people from the following organizations:

- Aircraft Owners and Pilot Association
- Aircraft and aircraft engine manufacturers
- Air Line Pilots Association
- American Association of Airport Executives
- Airports Council International – N/A
- Airport management and operations personnel
- Air Transport Association
- Engineering/environmental consulting firms
- FAA regional airport certification personnel
- Flight Safety Foundation
- Humane Society of the United States
- National Bird Strike Committees from over 20 countries
- International Civil Aviation Organization (ICAO)
- National Transportation Safety Board
- State wildlife agencies
- University and private research facilities
- US Department of Agriculture, Wildlife Services
- US Department of Interior, Fish and Wildlife Service
- US Department of Defense (Air Force, Navy, Army)
- Wildlife management companies

In addition to corporate and organizational membership, the following personnel also attend.

- Airline personnel
- Aircraft owners
- Aircraft and aircraft engine manufacturers
- Airline pilots
- Airport operations personnel
- Land-use planners
- FAA and Transport Canada airport inspectors

- Military aviation groups
- University researchers
- Waste management operators
- Wildlife agencies
- Wildlife control specialists

During the February 24, 2009 Congressional Hearings on Flight 1549, John Ostrom, President of the BSC USA, testified that "a major concern is that there has never been a joint industry/government body established to either address or define the issue. There is no recognized metric or standard to judge whether conditions are improving or worsening, and there is no comprehensive industry/government plan to address the hazard to aircraft and human life." Mr. Ostrom also testified that, in 2008, the BSC USA identified the following major goals:

- Establish and facilitate additional forums for exchanging knowledge and information regarding the nature and management of wildlife hazards to aviation and best practices and innovative technologies in the aviation and wildlife management industries. Currently, Bird Strike Committees USA and Canada sponsor a joint annual meeting, which is attended by both aviation and wildlife management professionals. Aviation and wildlife conflicts are a global problem, and we are working with our international partners, the International Bird Strike Committee (IBSC), on the possibility of future collaborative initiatives. In addition to our annual meetings and working with IBSC, we want to organize regional aviation wildlife hazard workshops for airports and flying communities around the country. The purpose of these workshops will be to establish a grassroots education and awareness initiative on the hazards of wildlife to aviation and measures that can be taken to mitigate them.
- Serve as the liaison to national and International Bird Strike Committees and to other professional aviation and wildlife organizations. In this capacity, we will provide timely and informed advice on aviation wildlife hazard management issues to governmental agencies, decision makers and others

who are responsible for the nation's air transportation system.

- Bird Strike Committee USA is working with the Federal Aviation Administration to establish a formal relationship which identifies Bird Strike Committee USA as the national committee for aviation wildlife hazard issues in the context of the International Civil Aviation Organization (ICAO) Airport Services Manual, Part 3 – Bird Control and Reduction.
- Conduct and promote communications efforts to enhance the awareness of wildlife hazards to aviation and efforts to reduce that threat. It is our intention to ensure that everyone involved in or affected by aviation-wildlife conflicts are fully cognizant of the issues as well as the tools and procedures needed to mitigate threats. Working with governmental agencies and aviation industry organizations, we intend to develop an informational awareness campaign and communications plan for the aviation community through the use of promotional materials and aviation industry media outlets.
- Contribute to the public's understanding of wildlife hazard management and its significance to the safety of air travel. In the wake of the recent US Airways 1549 accident, it is even more critical that a safety and awareness campaign targeted to the general public be developed describing the positive work being done with the current airport wildlife management programs and their impact on safety of flight.
- Promote professionalism in wildlife management programs on airports through professional development of individuals working in aviation wildlife hazard management. Managing our nation's wildlife resources at airports can be controversial, and it is our intent that persons performing this work be fully trained and knowledgeable of the tools available and of best practices. Bird Strike Committee USA is working to develop formal partnerships with industry organizations, academia and the private sector to create future training programs and

ethical guidelines for personnel conducting wildlife hazard management activities on airports.

- Promote the collection and analysis of accurate wildlife strike data for military and civil aviation in the USA as a foundation for a) understanding the nature of strike hazards, b) developing effective and appropriate management programs, and c) evaluating the efficacy of management programs. Analysis of strikes from US airports and airlines indicated that less than twenty percent of all strikes were reported to the FAA. Bird Strike Committee USA supports more aggressive reporting of all wildlife strikes, up to and including making strike reporting mandatory. In addition to increased strike reporting, it is imperative that we develop some form of performance measurement indicators that clearly identify progress towards specific goals in reducing wildlife hazards to aviation. As a previous Steering Committee Member once said, "We can't achieve success if we don't know where we are going." Finally, we need to make all of the information in the FAA's National Wildlife Strike Database available to airports and industry professionals in order to foster increased collaboration amongst groups with similar problems to develop more effective management programs.
- Anticipate future wildlife challenges to aviation and provide leadership in promoting education, research and development of effective methods for reducing wildlife hazards to aviation. Today, there is no single "clearinghouse" where yesterday's problems and tomorrow's solutions for aviation/wildlife conflicts can be brought together. It is our goal that Bird Strike Committee USA provide the future forum for scientific discussions as well as operational testing and standards for measuring performance of tools and processes for an effective and comprehensive aviation wildlife management program.

In 2009, the 11th combined meeting of Bird Strike Committee USA and Bird Strike Committee Canada will be conducted. Previous separate meetings of the two organizations have occurred over the last decade in various locations throughout the United States and Canada.

Presentations have included papers, posters and demonstrations on wildlife control techniques, new technologies, land-use issues, training, engineering standards, and habitat management. These presentations have all contributed to an enhanced awareness of the bird hazard issue.

The International Bird Strike Committee

The **International Bird Strike Committee (IBSC)** is a voluntary association of representatives from organizations who aim to improve commercial, military, and private aviation flight safety, by sharing knowledge concerning the reduction of the frequency and risk of collisions between aircraft and birds or other wildlife. The IBSC facilitates the following: collection, analysis, and dissemination of data to describe and define operational as well as functional, regulatory, and legal aspects of the bird strike risk to aviation; the description and evaluation of methods to define and reduce the severity, frequency and costs of bird strikes, to define and increase the ability of aircraft to tolerate the bird strike event, and to help air crews anticipate and react adequately; and cooperation and collaboration on investigative efforts in order to broaden the applicability of results while minimizing duplication of effort.

German Bird Strike Committee

In Germany, an organization called the German Bird Strike Committee (GBSC) was founded on 30 July 1964 at the suggestion of the Federal German Transport Minister, as a loose federation of the institutions and organizations involved in aviation. On 1 January 1981, the GBSC was entered into the German Register of Associations as a non-profit organization. Since 28 January 1998, the GBSC has been acting on behalf of the German government, its duties and targets being derived from Regulations and Decrees of Federal Ministries. The GBSC is a German organization for research on flight safety hazards arising from the living environment. Just as the name suggests, the prevention of aircraft/bird collisions is its primary duty.

The GSBC performs the following functions: advises associated airports (members of the GBSC), as well as national, federal and other authorities involved in aviation, on habitat management and bird strike prevention at and around airports; prepares and updates ecological reports on airports and their surroundings; drafts reports on changes in the environment of airports and their relevance to the bird strike situation; collects bird strike reports in Germany; compiles, updates and analyzes bird strike statistics for Germany; analyzes bird strike reports in order to gain a better understanding of the hazards posed by bird strikes; proposes, organizes and monitors research projects and experiments; offers training courses for technical and operational staff; cooperates with international organizations concerned with flight safety and maintains contacts with scientific institutions, environmental authorities and organizations, and other national bodies; evaluates national and international scientific literature; and publishes the scientific journal *Bird and Aviation.*

Australian Aviation Wildlife Hazard Group

The Australian Aviation Wildlife Hazard Group (AAWHG) has come about as a result of one of the recommendations of the Australian Transport Safety Bureau Report entitled "The Hazards Posed to Aircraft by Birds". The AAWHG was formed to enhance awareness of the safety issues surrounding bird and animal strikes; to provide an opportunity for bird and animal strike information, knowledge and advice to be shared; and to determine directions for future research, regulations and procedures to mitigate the risk posed by birds and animals to aircraft.

The International Center for the Study of Bird Migration in Latrun, Israel

Positioned directly on the path of the Western bird migration route, between Jerusalem and Tel Aviv, through which 500 million birds migrate in the spring and fall, the International Center for the Study of Bird Migration (ICSBM) is located at the Armored Corps Memorial Center, which attracts some 400,000 visitors annually. The center was founded by Dr. Yossi Leshem who was the Executive Director of the SPNI until 1995, and is a Senior Researcher in the Department of Zoology in the Faculty of Life Sciences at Tel Aviv University. ICSBM-Latrun is the joint initiative of Tel Aviv University and the Society for the Protection of Nature in Israel.

The center deals with the following projects:

- **Research:** Research projects include tracking storks, pelicans, vultures and cranes by attaching satellite transmitters to them; using raptors (e.g., barn owls and kestrels) as biological pest controllers of rodents; research of bird migration using bird radars by military (i.e., Navy and Air Force).
- **Flight Safety:** Through joint research with the Israeli Air Force, the number of accidents due to bird strikes was reduced by 76% since 1984, which saved the Israeli government 700 million US dollars (not to mention pilots' lives). A team of new immigrant Russian scientists manage the bird radar set at Latrun and transfer their bird migration findings to the Israel Air Force.
- **Education:** A multidisciplinary program was developed online (www.birds.org.il) under the title "*Migrating Birds Know No Boundaries*" and has been translated for use by more than 300 schools in Israel, 30 schools in the Palestinian National Authority and 30 schools in Jordan. **Environmental preservation and ecotourism:** The Center bears special importance in establishing Israel as a world leader in the area of protecting migrating birds and their habitats. It also represents a national focus in developing a network of birdwatching stations that will be based on research stations and field education, and exposing tourists and the Israeli public to the subject of ornithology.
- **International Cooperation: The Center is** currently collaborating with organizations from the Palestine Authority and from Jordan for the purpose of promoting mutual ambitions. The Center is also collaborating with research institutions and nature reserves in Europe and North America, and with international organizations such as UNESCO, for the purpose of declaring the Great Rift Valley "A World Heritage Site".

The UK Bird Strike Committee (UKBSC)

According to their website, the objectives of the UKBSC are as follows: to provide a forum for stakeholders to discuss bird/wildlife strike hazards and methods for reducing the associated risks; to provide a focal point and interface for interested organizations to discuss research and development issues and other means for reducing the hazard posed by birds/wildlife; to provide a consultative forum for the regulator when proposing changes to reporting, requirements or amending standards and guidance in dealing with airport bird/wildlife control; to act as an information source and exchange for those persons and organizations with a vested interest and expert knowledge in the subject matter; jointly in partnership with experts, to discuss issues affecting UK aerodrome license holders to assist in the evaluation of bird/wildlife hazards on, or in the vicinity of aerodromes; where appropriate, to encourage aerodrome license holders to provide training to National Occupational Standards and to ensure continuous competence of personnel engaged in aerodrome bird/wildlife control duties; to encourage the monitoring and recording of bird/wildlife activity and strikes on, and in the vicinity of, licensed aerodromes, by competent personnel, in accordance with Bird Control Management Plans; to encourage pilots, airline and aerodrome operators, ATC and others, to report all bird/wildlife strikes through appropriate reporting channels to the CAA, in accordance with the mandatory reporting requirements of Article 143 of the ANO and the guidance contained within CAP 772; to review national and international guidance (e.g., ICAO, EASA, FAA, Transport Canada and CASA) on bird/wildlife strike matters and encourage best practice guidance to be adopted and applied in the UK through CAA and other representative organization publications and communications.

The USAF Bird/Wildlife Aircraft Strike Hazard (BASH) Team

The BASH Team's goal is the preservation of war fighting capabilities through the reduction of wildlife hazards to aircraft operations. The BASH team is the Air Force's point-of-contact for worldwide on-site technical assistance. BASH is also responsible for developing research programs to reduce bird strike potential around airfields and during low-level flight operations. The team manages a voluminous aircraft bird strike database. The database is often used for conducting detailed statistical analysis for aircraft component design and environmental assessments.

FAA and USDA

The following is taken from a Memorandum of Understanding (MOU) between the FAA and USDA published June 27, 2005. The MOU continues the cooperation between the FAA and Wildlife Services (WS) for mitigating wildlife hazards to aviation. The FAA has the broad authority to regulate and develop civil aviation in the United States. The FAA may issue Airport Operating Certificates to airports serving certain air carrier aircraft. Issuance of an Airport Operating Certificate indicates that the airport meets the requirements of Title 14, Code of Federal Regulations, part 139 (14 CFR 139) for conducting certain air carrier operations. The WS has the authority to enter agreements with states, local jurisdictions, individuals, public and private agencies, organizations, and institutions for the control of nuisance wildlife. The WS also has the authority to charge for services provided under such agreements and to deposit the funds collected into the accounts that incur the costs.

14 CFR 139.337 requires the holder of an Airport Operating Certificate (certificate holder) to conduct a wildlife hazard assessment (WHA) when specific events occur on or near the airport. A wildlife management biologist who has professional training and/or experience in wildlife hazard management at airports, or someone working under the direct supervision of such an individual, must conduct the WHA required by 14 CFR 139.337. The FAA reviews all WHAs to determine if the certificate holder must develop and implement a wildlife hazard management plan (WHMP) that is designed to mitigate wildlife hazards to aviation on or near the airport. These regulations also require airport personnel implementing an FAA-approved WHMP to receive training conducted by a qualified wildlife damage management biologist.

The FAA and the WS agree that the WS has the professional expertise, airport experience, and training to provide support to assess and reduce wildlife hazards to aviation on and near airports. The WS can also provide the necessary training to airport personnel. Most airports lack the technical expertise to identify underlying causes of wildlife hazard problems. They can control many of their wildlife problems by following proper instruction in control techniques and wildlife species identification from qualified wildlife management biologists. Situations arise where control of hazardous wildlife is necessary on and off airport property (i.e., roost relocations, reductions in nesting populations, and removal of wildlife). This often requires the specialized technical support of WS personnel. The FAA or the certificate holder may seek technical support from WS to lessen wildlife hazards. This help may include, but is not limited to, conducting site visits and WHAs to identify hazardous wildlife, their daily and seasonal movement patterns, and their habitat requirements. WS personnel may also provide the following services: support with developing WHMPs, including recommendations on control and habitat management methods designed to minimize the presence of hazardous wildlife on or near the airport; training in wildlife species identification and the use of control devices; support with managing hazardous wildlife and associated habitats; and recommendations on the scope of further studies necessary to identify and minimize wildlife hazards.

Unless specifically requested by the certificate holder, WS is not liable or responsible for development, approval, or implementation of a WHMP required by 14 CFR 139.337. Development of a WHMP is the responsibility of the certificate holder. The certificate holder will use the information developed by WS from site visits and/or conducting WHA in the preparation of a WHMP. The FAA and WS agree to meet at least yearly to review the MOU, identify problems, exchange information on new control methods, identify research needs, and prioritize program needs. The WS personnel advise the certificate holder of their responsibilities to secure necessary permits and/or licenses for control of wildlife. This ensures all wildlife damage control activities are conducted under applicable federal, state, and local laws and regulations. The MOU defines, in general terms, the basis on which the parties will cooperate and does not constitute a financial obligation to serve as a basis for expenditures.

FAA, USAF, US Army, US EPA, US Fish and Wildlife Service & the USDA

The following is taken from a Memorandum of Agreement (MOA) published in 2003 between the agencies in the title of this Section to coordinate their missions and thereby more effectively address existing and future environmental conditions contributing to aircraft wildlife strikes throughout the United States. These efforts are intended to minimize wildlife risks to aviation and human safety, while protecting the nation's valuable environmental resources.

The agencies know the risks that aircraft wildlife strikes pose to aviation. The MOA acknowledges each agency's respective missions. Through the MOA, the agencies establish procedures necessary to coordinate their missions to more effectively address existing and future environmental conditions contributing to aircraft wildlife strikes throughout the United States. These efforts are intended to minimize wildlife risks to aviation and human safety, while protecting the nation's valuable environmental resources.

Aircraft wildlife strikes are the second leading cause of aviation-related fatalities. Globally, these strikes have killed over 400 people and destroyed more than 420 aircraft. While these extreme events are rare in comparison to the millions of annual aircraft operations, the potential for catastrophic loss of human life resulting from one incident is substantial. Approximately 97% of the reported civilian aircraft wildlife strikes involved common, large-bodied birds or large flocks of small birds. Almost 70% of these events involved gulls, waterfowl, and raptors. About 90% of aircraft wildlife strikes occur on or near airports, when aircraft are below altitudes of 2000 feet. Aircraft wildlife strikes at these elevations are especially dangerous because aircraft arc moving at high speeds and are close to, or on, the ground. At these times, aircrews are intently focused on complex takeoff or landing procedures and monitoring the movements of other aircraft in the airport vicinity. Aircrew attention to these activities while at low altitudes often compromises their ability to successfully recover from unexpected collisions with wildlife and to deal with rapidly changing flight procedures. As a result, crews have minimal time and space to recover from aircraft wildlife strikes.

Increasing bird and wildlife populations in urban and suburban areas near airports contribute to escalating aircraft wildlife strike rates. Experts from FAA, USAF, and Wildlife Services (WS) expect the risks, frequencies, and potential severities of aircraft wildlife strikes to increase during the next decade as the numbers of civilian and military aircraft operations grow to meet expanding transportation and military demands. 'Hazardous wildlife' are those animals, identified to species and listed in FAA and USAF databases, that are most often involved in aircraft wildlife strikes. Many of these species frequently inhabit areas on or near airports, cause structural damage to airport facilities, or attract other wildlife that pose an aircraft wildlife strike hazard. The databases are not intended to represent the universe of species concerning the signatory agencies, since more than 50 percent of the aircraft wildlife strikes reported to FAA or the USAF did not identify the species involved.

Not all habitat types attract hazardous wildlife. The agencies, during their consultative or decision-making activities, will inform regional and local land-use authorities of this MOA's purpose. The agencies will consider regional, local, and site-specific factors (e.g., geographic setting and/or ecological concerns) when conducting these activities and will work cooperatively with the authorities as they develop and implement local land-use programs under their respective jurisdictions. The agencies will encourage these stakeholders to develop land uses that do not attract hazardous wildlife. Conversely, the agencies will promote the establishment of land uses attractive to hazardous wildlife outside those criteria.

Wetlands provide many important ecological functions and values, including fish and wildlife habitats; flood protection; shoreline erosion control; water quality improvement; and recreational, educational, and research opportunities. To protect jurisdictional wetlands, Section 404 of the Clean Water Act (CWA) establishes a program to regulate dredge and/or fill activities in these wetlands and navigable waters. The agencies agree to work with landowners and communities to encourage and support wetland restoration or enhancement efforts that do not increase aircraft wildlife strike potentials.

The US Army Corps of Engineers (ACOE) has expertise in protecting and managing jurisdictional wetlands and their associated wildlife; the US Environmental Protection Agency (EPA) has expertise in protecting environmental resources; and the US Fish and Wildlife Service (USFWS) has expertise in protecting and managing wildlife and their habitats, including migratory birds and wetlands. Appropriate agencies will cooperatively review proposals to develop or expand wetland mitigation sites, or wildlife refuges that may attract hazardous wildlife.

During initial airport planning efforts, these agencies consult with airport proponents. As appropriate, the FAA or USAF will initiate signatory agency participation in these efforts. When evaluating proposals to build new civilian or military aviation facilities or to expand existing ones, the FAA or the USAF will work with appropriate signatory agencies to diligently evaluate alternatives that may help to avoid adverse effects on wetlands, other aquatic resources, and federal wildlife refuges. If these or other habitats support hazardous wildlife and there is no practicable alternative location for the proposed aviation project, then the appropriate signatory agencies, consistent with applicable laws, regulations, and policies, will develop mutually acceptable measures to protect aviation safety and mitigate any unavoidable wildlife impacts.

A variety of other land uses (e.g., storm water management facilities, wastewater treatment systems, landfills, golf courses, parks, agricultural or aquacultural facilities, and landscapes) attract hazardous wildlife and are, therefore, normally incompatible with airports.

The FAA, USAF, and WS personnel have the expertise necessary to determine the aircraft wildlife strike potentials of various land uses. When there is disagreement among the agencies about a particular land use and its potential to attract hazardous wildlife, the FAA, USAF, or WS will prepare a wildlife hazard assessment (WHA). Then, the appropriate signatory agencies will meet at the local level to review the assessment. At a minimum, that assessment will identify each species causing the aviation hazard, its seasonal and daily populations, and the population's local movements; discuss locations and features on and near the airport or land use attractive to hazardous wildlife; and evaluate the extent of the wildlife hazard to aviation.

When a potential wildlife hazard is identified, the agencies cooperate with the airport operator to develop a specific wildlife hazard management plan for a given location. The plan will meet applicable FAA, USAF, and other relevant requirements. In developing the plan, the appropriate agencies will use their expertise and attempt to integrate their respective programmatic responsibilities, while complying with existing laws, regulations, and policies. The plan should avoid adverse impacts to wildlife populations, wetlands, or other sensitive habitats to the maximum extent practical. Unavoidable impacts resulting from implementing the plan will be fully compensated pursuant to all applicable federal laws, regulations, and policies.

Whenever a significant aircraft wildlife strike occurs, or a potential for one is identified, any agency may initiate actions with other appropriate signatory agencies to evaluate the situation and develop mutually acceptable solutions to reduce the identified strike probability. The agencies will work cooperatively, preferably at the local level, to determine the causes of the strike and what can and should be done at the airport, or in its vicinity, to reduce potential strikes involving that species.

Information and analyses relating to mitigation that could cause or contribute to aircraft wildlife strikes should, whenever possible, be included in documents prepared to satisfy the National Environmental Policy Act (NEPA). This should be done in coordination with appropriate agencies to inform public and federal decision makers about important ecological factors that may affect aviation. The agencies will cooperatively develop mutually acceptable and consistent guidance, manuals, or procedures addressing the management of habitats attractive to hazardous wildlife.

The MOA does not alter or modify compliance with any federal law, regulation or guidance (e.g., Clean Water Act; Endangered Species Act; Migratory Bird Treaty Act; National Environmental Policy Act; North American Wetlands Conservation Act; Safe Drinking Water Act; or the "no-net-loss" policy for wetland protection). The agencies will employ this MOA in concert with the federal guidance addressing wetland mitigation banking. The MOA is intended to improve the internal management of the executive branch to address conflicts between aviation safety and wildlife.

Despite all the work that government and private organizations are doing to fix the aircraft bird strike problem, no solution has been discovered and implemented yet. The final chapters of this book describe how several existing technologies can be integrated into a bird strike warning system that sends bird targets directly to the cockpit. This is the right solution, and it comes at the right time – all that is needed for implementation is government financial and regulatory support.

Chapter 7:

Controlling Birds at Airports

It's been over 36 years since the NTSB made its first recommendation to the FAA on how to prevent bird strikes, and the solution somehow has evaded us. When aircraft first started flying, they were slower than today's high-speed jets, and birds usually had time to get out of the way. Now, jet speeds approach 550 mph (or 230-287 mph below 10000 feet). This chapter explains the laws and federal agencies that protect birds and the methods used to control birds at airports. Much of the data in this chapter is taken from an FAA report titled "*Wildlife Hazard Management at Airports*" by Edward Cleary and Dr. Richard Dolbeer and a 1994 book entitled *Prevention and Control of Wildlife Damage*, a cooperation between Cornell, Clemson, University of Nebraska-Lincoln and Utah State University. You may also want to reference *The Peterson Field Guide to Birds of North America* to better acquaint yourself with the birds under discussion. After reading this chapter, you will gain an appreciation of the amount of time government agencies and wildlife personnel spend trying to prevent bird strikes – it's a 24/7/365 job!

Federal Agencies and Laws that Protect Birds

The US Department of Agriculture/Wildlife Services (USDA/

WS) provides federal leadership in managing problems caused by wildlife and has primary responsibility for responding to threats caused by migratory birds. According to the Memoranda of Understanding with the FAA and Department of Defense as well as guidelines published elsewhere, the USDA/WS assists federal, state, and local agencies, airport managers, the aviation industry and the military in reducing wildlife hazards on and in the vicinity of airports and air bases. In addition, it is the responsibility of USDA/WS personnel that observe existing or potential wildlife hazards at airports or air bases to immediately notify the appropriate aviation authorities. USDA/WS may enter into cooperative agreements to develop Wildlife Hazard Assessments (WHAs) and Wildlife Hazard Management Plans (WHMPs) and to conduct direct wildlife hazard reduction programs. USDA/WS biologists may provide training for airport and air base personnel in wildlife and hazard identification and the safe and proper use of wildlife control equipment and techniques. They may also provide recommendations and assistance to airport managers and air base commanders in obtaining federal, state, and local permits to remove protected wildlife species.

The US Air Force (USAF) Bird Aircraft Strike Hazard (BASH) Team at HQ Air Force Safety Center, Kirtland Air Force Base, New Mexico oversees the USAF wildlife strike reduction efforts. The BASH team maintains a wildlife strike database for strikes involving USAF aircraft, similar to the database maintained by the FAA for civil aircraft. Anytime bird habitats are modified by excavating or use of filling material, the airport must obtain a permit from the US Army Corps of Engineers (USACE), which is in charge of a wide range of functions related to water resources, among which is the protection of navigation and safeguarding the nation's water resources.

The US Environmental Protection Agency (EPA) was formed to safeguard the nation's environment. Its functions include setting and enforcing environmental standards and regulations related to air and water pollution, hazardous wastes, pesticides, and toxic substances. Their mission is accomplished through partnerships with state and local governments. EPA responsibilities include pesticide registration and regulation as well as the siting and construction of wastewater treatment and solid waste disposal facilities, which are permitted through state and local agencies. The FAA and USDA/WS may be consulted by airport authorities or state and local agencies to review impacts of proposed EPA-regulated projects on aviation safety. Approval or disapproval of a landfill site is the responsibility of the EPA, state and local governing bodies, and zoning boards. Before any pesticide may be used, it must be registered with the EPA and with the appropriate state pesticide regulating authority.

The mission of the US Fish and Wildlife Service (USFWS) is to conserve, protect, and enhance the nation's fish and wildlife and their habitats for the continuing benefit of all people. Responsible for the conservation and enhancement of migratory birds, threatened and endangered species, certain marine mammals, migrating fishes, and wetlands, the USFWS manages the National Wildlife Refuge System, enforces federal wildlife laws, and conducts biological reviews of the environmental impacts of development projects. The organization renders biological opinions on proposed federal activities that might impact federally listed or proposed endangered or threatened bird species or result in the destruction or adverse modification of designated or proposed critical habitats.

Air traffic controllers are federal employees and must report any unsafe conditions, including birds, to the appropriate airport personnel any time they are observed. The controllers are also required to issue advisory information on pilot-reported, tower-reported, or radar-observed and pilot-verified bird activity and relay bird activity information to adjacent facilities and to Flight Service Stations (FSS) whenever it appears that a bird hazard will become a factor in the area.

Most bird species are protected by federal and state laws, and the legal status of the birds needs to be checked before an attempt is made to control them. Migratory birds are protected under the Migratory Bird Treaty Act (MBTA) of 1918. The MBTA prohibits the taking of any migratory bird or their parts (including feathers, nests, and eggs). Under the MBTA, to "take" is defined as "to pursue, hunt, shoot, wound, kill, trap, capture, or collect, or any attempt to carry out these activities." "Take" does not, however, include habitat destruction or alteration, as long as there is not a direct taking of birds, nests, eggs, or parts thereof. Some species are further protected by the Endangered Species Act (ESA) of 1973, Section 9 of which prohibits any person from "taking" an endangered bird species. Under ESA, "take" is defined, in part, as killing, harming, or harassing. Under federal regulations, "take" is further defined to include habitat modification or degradation that actually results in death or injury to birds by significantly impairing essential behavioral patterns, including breeding, feeding, or sheltering.

Controlling the Bird Problem at Airports

Every airport has a bird problem; and for each of them, it is different. As a result of the airliner in the Hudson, the FAA is working on a rule that will require all commercial airports to conduct a wildlife hazard assessment (WHA). Consequently, more airports will be required to work harder to control their bird populations. In order achieve this sort of control, though, it is necessary to first examine what attracts birds to airports; the major attractants are food, water and cover. The vertical threat area extends from surface to 3000 feet because, according to the FAA bird strike database, 92% of bird strikes occur within these altitudes. The horizontal threat area comes from an FAA recommendation of 5 statute miles between the farthest edge of the airport and bird attractants if the attractant could cause hazardous wildlife movement into or across the approach or departure airspace. This distance comes from Advisory Circular 150/5200-33 (Hazardous Wildlife Attractants on or Near Airports).

Title 14 of the Code of Federal Regulations (Title 14 CFR) Part 139.337 currently requires that commercial airports in the United States conduct a Wildlife Hazard Assessment (WHA) when any of the following trigger events occurs at their airport: an air carrier aircraft experiences multiple wildlife strikes; an air carrier aircraft experiences substantial damage from striking wildlife; or an air carrier aircraft experiences an engine ingestion of wildlife. If any of these events occur, an airport must conduct a WHA and submit it to the FAA. The FAA then determines whether a Wildlife Hazard Management Plan (WHMP) is necessary. The WHA is conducted by a wildlife damage management biologist (often from the USDA) and provides the scientific basis for the development, implementation, and refinement of a WHMP. The wildlife species observed must be identified; and their numbers, locations, local movements, and daily and seasonal occurrences must be reported. In most cases, this requirement dictates that a 12-month assessment be conducted so that the seasonal patterns of birds and other wildlife using the airport and surrounding area during an annual cycle can be properly documented. Most regions of the United States have dramatic seasonal differences in terms of the numbers and species of migratory birds. Even for non-migratory birds like resident Canada geese, behavior and movement patterns can change significantly from season to season. If, after submitting the WHA, the FAA determines that a WHMP is required, then the WHMP must accomplish the following tasks: identify and provide information on hazardous wildlife attractants on or near the airport; identify appropriate wildlife management techniques to minimize the wildlife hazard; prioritize appropriate management measures; recommend necessary equipment and supplies; identify training requirements for the airport personnel who will implement the WHMP; and identify when and how the plan will be reviewed and updated. For a complete description of the process that airports need to go through for commercial certification, see Appendix 6.

According to the Wildlife Hazard Management at Airports Report (Cleary & Dolbeer), there are four ways to mitigate the threat posed by birds at an airport: (1) aircraft flight schedule modification; (2) habitat modification and exclusion; (3) repellent and harassment techniques; and (4) wildlife removal. These methods must be integrated into the WHMP as appropriate.

Aircraft Flight Schedule Modification

Although not generally practical for regularly scheduled commercial traffic on larger airports, there may be various situations when flight schedules of some aircraft can be adjusted to minimize the chance of a strike with a bird species that has a predictable pattern of movement. For example, pilots could be advised not to depart during a 20-minute period at sunrise or sunset during winter when large flocks of blackbirds cross an airport going to and from an off-airport roosting site. I have actually experienced this sort of adjustment in my military flight training schedule. While I was learning to be a T-38 instructor pilot at Randolf Air Force Base (AFB), TX. I remember a restriction that prevented us from flying at night. A place called Bracken Cave is located just outside of San Antonio and has over 20 million bats, the largest concentration of bats in the world. A group called Bat Conservation International is dedicated to protecting the cave and the land around it. I received no night instruction as a pilot because of that cave, and then, as luck would have it, after I passed my check rides and became certified as an instructor pilot, my first night flight was with a student in formation! In places such as Midway Atoll, where albatrosses and other seabirds are abundant during parts of the year, scheduling nighttime arrivals and departures (i.e., those times when birds are not flying) might be the only means of avoiding strikes. And finally, air traffic controllers, acting on advice by the wildlife specialists, will actually close a runway with unusually high bird activity or until wildlife control personnel can disperse them.

Habitat Modification and Exclusion

Another one of the control techniques is to modify the bird's environment or habitat. The goal of this technique is to make the environment undesirable or unavailable, so the elements of food, water and cover are removed. Some habitat modifications include elimination of standing water, removal or thinning of trees and removal of brush, and management of grass height. Buildings can also be modified to reduce or eliminate sites for roosting or nesting. FAA Order 5200.5A provides guidelines for the establishment, elimination, or monitoring of landfills, open dumps, or waste disposal sites on or in the vicinity of airports.

Bird attractants are defined more precisely in the book *Prevention and Control of Wildlife Damage*: "Most airports support an abundance and variety of foods, such as seeds, berries, grass, insects, grubs, earthworms, small birds, and small mammals. Seeds and berries are sought by several migratory and resident birds, such as sparrows, finches, starlings, blackbirds, mourning doves, common pigeons, and waterfowl. Geese are attracted to open expanses of grasses. Gulls, starlings, robins, and crows often feed on earthworms on the surface of the ground following a rain. Gulls are opportunistic feeders and frequently feed on grasshoppers and ground-nesting birds. Raptors are attracted to airports because of rodents, birds, and other small animals that are harbored by tall, poorly maintained grass stands and borders."

Another bird attractant is garbage dumps. Landfills are often located on or near airports because both are often built on publicly owned lands. Landfills contribute to bird strike hazards by providing food sources and loafing areas that attract and support thousands of gulls, starlings, pigeons, and other species. Landfills are a major attraction for gulls, the most common bird involved in bird strikes. Waste paper, paper bags, and other litter blowing across the ground attract gulls because it is mistaken for other gulls or for food. A gull that is attracted to litter lures other gulls and encourages flocking.

Water acts as a magnet for birds. Birds of all types are drawn to open water for drinking, bathing, feeding, loafing, roosting, and protection. All standing water on an airport should be eliminated to the greatest extent possible. Standing water areas should be filled or modified to allow rapid drainage of depressions that accumulate standing water after rain. This is particularly important at coastal airports, where fresh water is highly attractive to birds for drinking and bathing. Rainy periods provide temporary water pools at many airports. Many airports have permanent bodies of water near or between runways for landscaping, flood control, or wastewater purposes. These permanent sources of water provide a variety of bird foods, including small fish, tadpoles, frogs, insect larvae, other invertebrates, and edible aquatic plants. Temporary and permanent waters, including ponds, borrow pits, sumps, swamps, and lakes attract gulls, waterfowl, shorebirds, and marsh birds.

Birds need cover for loafing, roosting, resting and nesting. Trees, brushy areas, weed patches, shrubs, and airport structures often provide suitable habitats to meet these requirements. Almost any area that is free from human disturbance may provide a suitable roosting site for one or more species of birds. Starlings, pigeons, house sparrows, and swallows often roost or nest in large numbers in airport buildings or nearby trees, shrubs, or hedges. Large concentrations of blackbirds and starlings are attracted to woody thickets for winter roosting cover. Gulls often find safety on or near runways of coastal airports when storms prevent their roosting at sea, on islands, or on coastal bays. Non-migratory Canada geese in urban areas, if left undisturbed, will establish territories on corporate lawns, golf courses, and even building roofs associated with nearby ponds. Pigeons, house sparrows, and European starlings use building ledges, abandoned buildings, open girders and bridge work, and dense vegetation for cover. Gulls and other birds concentrate at locations where people regularly provide food, such as bread and seeds. Feeding birds should be prohibited both on, and in the vicinity of, airports. Blackbirds use marsh vegetation, such as cattails, for nesting and roosting. Many bird problems can be solved by eliminating the availability of such areas either through removal or by exclusion. When selecting and spacing plants for airport landscaping, plants that produce fruits and seeds desired by birds should be avoided. The creation of areas of dense cover for roosting, especially by European starlings and blackbirds, should be avoided. Thinning the canopy of trees, or selectively removing trees to increase their spacing, can help eliminate bird roosts that tend to form in trees on airports.

Many airports are located along traditional annual bird migration routes, in which case birds may suddenly appear in large flocks on or over an airport on their annual migration, even when the airport itself offers no particular attraction. Dates of migration vary by species and area. Flock size of a given species may vary widely from year to year depending on time of year, weather conditions, and many other factors. Shorebirds, waterfowl, gulls, and other birds often make daily flights across airports from their feeding, roosting, nesting, and loafing areas. Airports near cities may experience early morning and late afternoon roosting or feeding flights of thousands of starlings.

If food, water, or cover cannot be eliminated by habitat modification, then actions can sometimes be taken to exclude the wildlife from the desired resource. Exclusion involves the use of physical barriers to deny birds access to a particular area. As with habitat modification, exclusion techniques, such as installing a covered drainage ditch instead of an open ditch, can initially be costly; however, exclusion provides a permanent solution that is not only environmentally friendly, but when amortized over many years, might actually be the least expensive solution. Physical exclusion devices include spike strips, netting, floating deterrents, doors, and other devices that deter or prohibit perching or entry. Though these devices can be effective on a small scale, they are often cost-prohibitive on a large scale.

Prevention is often the easiest solution. In aircraft hangars and storage facilities, making sure that doors are closed whenever not in use will prevent some problem situations and is cost-effective. Exclusion devices should be installed when birds are not occupying the area to be excluded. If it is impossible to remove all birds at one time, one-way doors can be used to prevent re-entry of the persistent birds after they finally leave to find food or water. Maintenance of exclusion material (e.g., repairing holes in netting) is essential for maintaining effectiveness. Heavy plastic strips hung vertically in open doorways have been successful in some situations to exclude birds (Johnson & Glahn 1994). These strips allow quick entry for humans or machinery while keeping birds out.

Exclusion can also be achieved by removing the attractant. While removing a hangar is not feasible or practical, removal of trees to prevent roosting may work. For airports, roosts present several problems: (1) they attract large numbers of birds to the airport; (2) feces build up and can become slick when wet, unsightly, and contain various diseases; and (3) they present an auditory nuisance to passengers. If tree removal is not feasible, then thinning a tree's branches makes it less attractive for roosting since the tree no longer holds heat effectively. Pruning every third limb is the recommended pruning strategy for trees that host bird roosts. Some birds can be excluded from ponds or other areas by using overhead wire grids (Fairaizl 1992, Lowney 1993). These lines should be made visible to the birds by hanging streamers or other objects at intervals along the wires. The objective is to discourage bird loafing and feeding activities, not to cause bird injury or death. Overhead wire networks generally require little maintenance other than maintaining proper wire tension and replacing an occasional broken wire. The spacing of the wires varies with the species being excluded.

During the design phase of buildings, hangars, bridges, and other structures at airports, architects can consult biologists to minimize exposed areas that birds can use for perching and nesting. For example, tubular steel beams are much less attractive as perching sites for starlings and pigeons than I-beams. If desirable perching sites are present in older structures, then access to these sites (e.g., rafters and girded areas in hangars, warehouses, and under bridges) can often be eliminated with netting. Curtains made of heavy-duty plastic sheeting, cut into 12-inch strips, and hung in warehouse or hangar doorways, can discourage birds from entering these openings. Anti-perching devices, such as spikes, can be installed on ledges, roof peaks, rafters, signs, posts, and other roosting and perching areas to keep certain birds from using them. Changing the angle of building ledges to 45 degrees or more will deter birds from settling there. It is emphasized, though, that incorporating bird exclusion or deterrence into the design of structures is the most effective, long-term solution.

Gull and waterfowl use of retention ponds and drainage ditches can be reduced with over-head wire systems. A system of wires spaced 10 feet apart or in a 10'x10' grid will discourage most gulls and waterfowl from landing. Similar wire systems have succeeded in keeping gulls off roofs and out of landfills and crows out of electrical substations. When it is desirable to eliminate all bird use, netting can be installed over small ponds and similar areas. However, birds sometimes get tangled in the netting, and maintenance problems arise with high winds and freezing weather. Complete coverage of ponds with plastic, 3-inch diameter "bird balls" or floating mats will completely exclude birds and yet allow evaporation of water. If steeply sloping ground is incorporated into the designs of ponds, then wading birds such as herons will be discouraged from gathering there. Use of culverts to totally cover water in drainage ditches is recommended whenever possible.

Repellant Techniques

Repellent and harassment techniques are designed to make the area or resource that is desired by wildlife unattractive or to make the wildlife uncomfortable or fearful. Over the long term, the cost-effectiveness of repelling birds usually does not compare favorably with habitat modification or exclusion techniques. No matter how many times birds are driven from an area that attracts them, they or other individuals of their species will return as long as the attractant is accessible. However, habitat modifications and exclusion techniques will never completely rid an airport of problem birds; therefore, repellent techniques are a key component of any wildlife hazard management plan.

Repellents work by affecting the birds' senses through chemical, auditory, or visual means. Acclimation of birds and mammals to most repellent devices or techniques is a major problem. When used repeatedly without added reinforcement, birds soon learn that the repellent devices or techniques are harmless; the devices become a part of their "background noise", and they ignore them. Critical factors to be recognized in deploying repellents are that there's no "silver bullet" that can solve all problems and no standard protocol or set of procedures that is best for all situations. Repelling wildlife is an art as much as a science. The most important factors are having motivated, trained, appropriately equipped personnel who understand the wildlife situation on their airport and recognizing that each wildlife species is unique and will often respond differently to various repellent techniques. Even within a group of closely related species, such as gulls, the various species will often respond differently to various repellent techniques. In the future, advances in electronics, remote sensing capabilities, and computers can be integrated to develop "intelligent" systems that can automatically deploy repellent devices (e.g., noisemakers, chemical sprays) when targeted birds enter a designated area. These devices might help reduce habituation and increase effectiveness of repellents in some situations. However, these devices will never replace the need for trained people on the ground to respond appropriately to incursions by a variety of highly adaptable wildlife species.

Chemical repellents, toxicants, and capturing agents must be registered with the EPA or Food and Drug Administration (FDA) before they can be used to manage birds on airports. Products must also be registered in each state. Prevention and Control of Wildlife Damage by Hygnstrom *et al.* (1994) provide a listing of chemical products, organized by active ingredient and by company name, registered for birds and mammals.

For application on perching structures, polybutenes are available in liquid or paste form. These sticky formulations make birds uncomfortable when they land or step on them, encouraging the birds to look elsewhere to perch or roost. To be effective, all perching surfaces in a problem area must be treated or the birds will merely move a short distance to an untreated surface. Under normal conditions, the effective life of these materials is 6 months to 1 year, but dusty environments can substantially reduce the life expectancy. Once the material loses effectiveness, it is necessary to remove the old material and apply a fresh coat. Applying the material over duct tape, rather than directly to the building ledge or rafter surface, will facilitate clean up.

There are two chemicals presently registered as bird repellents for turf (i.e., grass). One is an anthraquinone formulation for repelling geese, and the other is methyl anthranilate. Anthraquinone apparently acts as a conditioned-aversion repellent with birds. Birds ingesting food treated with anthraquinone become slightly ill and develop a post-ingestion aversion to the treated food source. Birds visually identify anthraquinone in the UV light spectrum and become conditioned to avoid the treated food source. Because of its conditioned-aversion properties, anthraquinone use does not require treatment of the entire airfield, but only areas where birds are grazing and/or higher risk areas, such as runway approaches. The other repellent, methyl anthranilate, is an artificial grape flavoring commonly used in foods and beverages. Birds have a taste aversion to methyl anthranilate, apparently reacting to it in much the same way that mammals react to concentrated ammonia (smelling salts). Methyl anthranilate is registered under formulations as a feeding repellent for geese and other birds on turf. Both anthraquinone and methyl anthranilate products are liquid formulations applied by sprayer to the vegetation. Effectiveness of these sprays in repelling geese can be variable, depending on growing conditions, rainfall, mowing, and availability of alternate feeding areas. In general, repellency based on conditioned aversion is longer lasting than repellency based on taste. Methyl anthranilate formulations are also available for application to pools of standing water on airports and at other locations to repel birds from drinking and bathing. This application is probably best for temporary pools of water after rainfall, where repellency of only a few days is needed. A methyl anthranilate formulation is also available for use in fogging machines (thermal or mechanical) to disperse birds from hangars, lawns, and other areas.

Scaring agent Avitrol (4-Aminopyridine) is registered for repelling pigeons, house sparrows, blackbirds, grackles, cowbirds, starlings, crows, and gulls from feeding, nesting, loafing, and roosting sites. Birds eating Avitrol-treated bait react with distress symptoms and calls, behaviors that frighten away other birds in the flock. Avitrol, although registered as a scaring agent, is lethal to the birds that eat treated bait. Therefore, Avitrol is a toxin to the birds that consume treated bait. Avitrol-treated bait is diluted with untreated bait, so most birds in the flock do not ingest treated bait. The primary use of Avitrol at airports has been in pigeon control around buildings. The use of Avitrol requires knowledge of the feeding patterns of the birds, proper pre-baiting procedures to ensure bait acceptance and avoidance of non-target species, and removal of dead birds after treatment.

Harrassment Techniques

Many visual and sound-making devices are commercially available for scaring birds. These include gas-operated exploders, pyrotechnics, electronic noisemakers, bird distress calls, standing or pop-up bird decoys, eyespot balloons, raptor models, strobe or flashing lights, reflective plates or lines and water spray devices. The value of these devices, however, is usually limited to short-term control. Although bird damage can sometimes be reduced by using only one type of scaring device, better results over longer periods are often achieved by using a combination of devices and/or by changing methods frequently. In addition, pieces of scaring equipment, and especially sound-making devices, are usually more effective when moved often; this is because birds will eventually ignore any scaring device that is left in the same place or that emits sound in the same regular pattern over a long period of time.

It is important to start the scaring regime before the birds establish regular feeding patterns at a site. Once regular habits are established, they are difficult to break by means of scaring techniques. Although the majority of birds may be scared away initially by scaring methods, some individuals will soon ignore the control methods. These "hard-to-scare" individuals attract others to the feeding site and require a control method involving real danger from the bird's point of view, such as pyrotechnics or exploders, reinforced by human presence. The effectiveness of scaring devices can be improved by incorporating the use of rifles or shotguns (with permit) to remove birds that have habituated. Because of all of the variables involved, the success of a scaring program is dependent on the skill and motivation of the operator. Scaring devices will not be effective unless used aggressively and as part of a carefully planned program.

Bird dispersal patrol teams can be used to harass birds in the immediate area of larger aquaculture facilities. Patrols must be adequately supplied with radio-equipped vehicles, bird distress calls, shotguns, live ammunition, and pyrotechnics. Patrol personnel must be trained in bird identification and dispersal methods. Blackbirds, cormorants, herons, and other species establish roosts, especially during winter, that include many individuals (or in the case of blackbirds, hundreds of thousands). These birds may cause significant losses if they feed in aquaculture facilities as a significant amount of fish will be eaten. Choosing the most effective combination of scaring devices requires careful consideration. One must match the devices to the bird species causing damage, assess the cost of the equipment and labor requirements, and consider possible interference with culture operations. For example, loud noises disturb spawning catfish, but they may also disturb neighbors or others near the aquaculture site.

The automatic exploder resembles a small cannon. It commonly operates on propane gas or acetylene and emits loud explosive blasts at adjustable time intervals. While the number of exploders needed will vary from site to site, one exploder can usually cover 3 to 5 acres if used properly and reinforced with other control techniques. Explosion frequency is important since short intervals increase the chance that birds will become accustomed to the sound. Timers that automatically start and stop the operation to produce irregular explosion intervals, and rotary mounts that change the direction of the sound after each explosion improve the device's effectiveness. For best results, exploders must be removed every 1 to 2 days to a different part of the facility. If necessary, the exploders should be elevated to prevent foliage or adjacent equipment from interfering with sound projection. Exploders have been effective for herons, egrets, cormorants, diving ducks, and blackbirds.

Harassment of birds can be accomplished by firing shellcrackers from a 12-gauge shotgun. These shells contain a firecracker that is projected 50 to 100 yards before exploding. Since wads from the shell may stick in the gun, it is important to check the barrel after each shot and regularly clean the gun. Breech opening, open-bore shotguns are required. Other pyrotechnic wildlife dispersal devices, variously known as noise, bird, clow, racket, or whistle bombs; noise rockets; or bird whistles, are among the most effective scaring devices. Though the range of these projectiles is only 35 to 75 yards, they are less expensive and more convenient to handle than shellcrackers. Possession and use of pyrotechnics may require a permit from the local, county, and/or state fire marshal. Blackbirds and grackles have been effectively frightened by .22 caliber birdshot.

Many species of birds emit calls that communicate alarm or distress to other birds of the same species, and broadcasted recordings of these calls can frighten and repel some bird species. Reaction to the calls varies with species of bird, location, size of area, and time of year. For best results, broadcast distress calls as birds begin to arrive. A timing device can be used to play calls at predetermined intervals. Lengthen the time between broadcast intervals as much as possible while still achieving the desired response. Birds habituate to distress calls if they are played frequently or over a long period in the same location. Calls need to be reinforced by other methods. Alarm calls have been used successfully on black-crowned night-herons, gulls, and blackbirds.

A variety of lights, including strobe, barricade, and revolving units, have been used to frighten birds, but with mixed results. Of these, strobe lights similar to those used on aircraft are most effective in scaring night-feeding birds. These extremely bright flashing lights have a blinding effect, causing confusion which reduces a bird's ability to catch fish. Black-crowned night-herons, however, may avoid the bright glare by landing with their backs to the lights or by moving to darker areas. Avoidance may be minimized by increasing the number of lights to cover the unprotected areas. Flashing amber barricade lights, such as those used at construction sites, and revolving or moving lights may also frighten birds when these units are placed on raceway walls or fish pond banks. Most birds, however, rapidly become accustomed to such lights, and their long-term effectiveness is questionable. In general, the type of light, the number of units, and their location are determined by the size of the area to be protected and by the power source available.

Water spray from rotating sprinklers placed at strategic locations in or around ponds or raceways will repel certain birds, particularly gulls (Svensson 1976). Individual birds may become accustomed to the spray, though, and actually feed among the sprinklers. Best results are obtained when sufficient water pressure is used and the sprinklers are operated on an on-off cycle. The sudden start-up noise also helps to frighten the birds.

Most visual repellents are simply a variation on an ancient theme – the scarecrow. In general, visual repellents, such as hawk silhouettes, balloons, flags, and Mylar reflecting tapes, have shown only short-term effectiveness and are inappropriate for use as a long-term solution to bird problems on airports. Most short-term success achieved with these devices is likely attributable to "new object reaction" rather than to any actual scaring effect produced by them. One visual deterrent that has been successfully used in recent years is the display of dead birds in a "death pose". Several experiments and field demonstrations have shown that a dead turkey vulture (freeze-dried taxidermy mount with wings spread), hung by its feet in a vulture roosting or perching area, will cause vultures to abandon the site. Initial trials using dead gulls and ravens suspended from poles have also shown promising results in dispersing these species from feeding and resting sites. The dead bird must be hung in a "death pose" to be effective; dead birds lying supine on the ground or in the roost are generally ignored or might even attract other birds. Permits must be in place before federally protected migratory birds can be obtained and used as "dead-bird deterrents". Research is under way to determine if artificial "dead birds" can be developed that will be just as effective as the taxidermy mounts.

Another new concept in visual repellency that has shown utility in recent years is the use of hand-held laser devices that project a 1-inch diameter red beam to disperse birds. These devices have been used successfully to disperse birds such as Canada geese, double-crested cormorants and crows from nighttime roosting areas in reservoirs and trees. Although the use of a laser to alter bird behavior was first introduced nearly 30 years ago (Lustick 1973), it received very little attention until recently when it was tested by the WS National Wildlife Research Center. Results have shown that several birds, including cormorants, waterfowl, gulls, vultures, and crows, exhibit avoidance of laser beams during field trails (Glahn *et al.* 2001, Blackwell *et al.* 2002). The repellent or dispersal effect of a laser is due to the intense and coherent mono-wavelength light that, when targeted at birds, may elicit changes in physiological processes (USDA APHIS 2001). Best results are achieved under low-light conditions (i.e., sunset through dawn) and when targeting structures or trees near roosting birds. Habituation to lasers has not been observed in field situations (USDA APHIS 2001). Extreme caution should be used when applying lasers around airfields to prevent interfering with aircraft operations. No damage to the avian eye has been recorded, as the avian eye filters most damaging radiation, unlike the human eye (USDA 2003). Advantages are effectiveness at long range (over ¼ mile) and lack of noise. Lasers have also shown some effectiveness in dispersing birds from hangars. Effectiveness is diminished or nonexistent in daylight conditions. As with the use of firearms, the use of lasers in an airport environment obviously requires caution.

Ultralight aircraft have been used to intercept large flocks of birds and herd them away from commercial facilities. This has been most effective with large concentrations of pelicans. Radio-controlled (RC) model aircraft, which provide both visual and auditory stimuli, occasionally have been used to harass birds on airports. One advantage is that the RC aircraft is under the control of a person and can be directed precisely to herd the birds away from the airport runway. A second advantage is that the RC aircraft can be deployed on an "as needed" basis with little maintenance needed between flights. Some RC aircraft have been designed to mimic the appearance of a falcon and even to remotely fire pyrotechnics.

The disadvantage is that a trained person is required to operate the RC aircraft in an airport environment. Before using RC aircraft, make sure that the radio frequencies used are compatible with other radio uses in the airfield environment. Ultralights, radio-controlled model airplanes and model raptors have been used but are expensive, subject to weather conditions (e.g., high winds), and in the case of ultralights, may place humans at risk. They also may be ineffective on species that seek safety by diving underwater. Interesting to note, US Patent Number 4964331 (April 1990) describes a RC aircraft that includes a receiver to control the launching of a special purpose cracker cartridge having a capsule which burns with a faint smoke trail for a predetermined period of time, after which the projected capsule explodes with a brilliant flash, loud noise, and a small cloud of smoke. By this technique, birds are actually chased out of the path of a runway at an airport. The cannon firing mechanism which launches the exploding capsule is controlled by a manually operated transmitter from the ground.

Falconry, the practice of using falcons and hawks to chase/hunt other wildlife species and then return to the handler, is regulated under both federal and state laws, and all raptors in the United States are protected under various statutes. Any "take" of a raptor must be done under the appropriate permit to be legal. The advantage of falconry is that the birds on the airport are exposed to a natural predator for which they have an innate fear. The disadvantages are that a falconry program is often expensive (Chamorro & Clavero 1994), requiring a number of birds that must be maintained and cared for by a crew of trained, highly motivated personnel. Effectiveness of falconry programs in actually reducing strikes has been difficult to evaluate. The care and housing of falcons can be expensive, and there are drawbacks to using falcons to disperse birds from damage or potential damage sites – the releasing of another "wildlife hazard" could result in a bird strike and liability (Hahn 1996)

Although the use of falconry at airports was evaluated thirty years ago (Blokpoel 1976), the following suggestions still remain valid: (1) properly trained birds of prey of the right species for the job at hand, used regularly and persistently by skilled and conscientious personnel, are effective in clearing birds from airfields during daylight and good weather; (2) for good results, daily operations on a year-round basis are required in most cases; (3) several falcons are required to have at least one bird ready at all times; and (4) to obtain, train, operate, and care for falcons, a staff of at least two full-time, well-trained personnel is required. The Port Authority of New York and New Jersey has a 5-year, $3 million contract with a company to do falconry at Kennedy Airport where falconry has been successfully used for a decade.

The use of trained dogs, especially border collies, to chase geese and other birds from golf courses, airports, and other sites is a recent development. Properly trained dogs can be effective at harassing birds and keeping them off turf and beaches (Conover & Chasko 1985, Woodruff & Green 1995). As with falcons, the advantage here is in exposing the birds to a natural predator. Dogs are particularly effective in harassing Canada geese. Around water, the use of trained dogs to scare birds appears most effective when the body of water to be patrolled is less than 2 acres in size (Swift 1998). Although dogs can be effective in keeping birds off individual properties, they do not contribute to a solution for the larger problem of overabundant bird populations (Castelli & Sleggs 1998). Likewise, the disadvantage is that the dog must be under the control of a trained person at all times, and the dog must be cared for and exercised 365 days a year. A dog will have little influence on birds that are flying over the airport. The use of dogs at Johannesburg and Durban International Airports in South Africa (2002) resulted in a significant reduction in the bird hazard risk and led to improved aviation safety

Wildlife Removal

Habitat modification, exclusion, and repellent techniques are the first lines of action in any WHMP. However, these actions will not solve every problem; therefore, hazardous wildlife sometimes must be removed from an airport. Such removal can be accomplished by capturing and relocating or by killing the birds. With few exceptions, a Federal Migratory Bird Depredation Permit, and in many cases a state permit, is required before any migratory birds may be taken (i.e., captured or killed). Any capturing or killing must be done humanely and only by people who are trained in wildlife species identification and the techniques to be deployed.

Relocation of wildlife often involves stress to the relocated animal, poor survival rates, and difficulties in adapting to new locations or habitats, or the animal may simply leave the area. However, relocation can be an effective method of reducing the risk of wildlife strikes when special-status species are involved. Relocation of sensitive wildlife, however, has been successfully implemented at some airports. This option normally includes habitat management, audio harassment, vehicular harassment, and harassment with pyrotechnics.

The major advantage of live trapping is selectivity and also the fact that any non-target birds can be released unharmed. The major disadvantage is that live trapping is often labor-intensive. Traps must be tended frequently to remove captured animals and, in the case of cage traps with decoy birds, to provide food and water. Trapping is used on some airports to remove raptors (i.e., hawks and owls) from the aircraft operating area. Because raptors are desirable components of bird communities, most permits for trapping raptors require that the birds be banded and relocated into suitable habitat at least 50 miles from the airport. Live trapping, using walk-in type traps on roofs or other isolated sites, can be used to remove pigeons at airports. Captured pigeons can only be euthanized if American Association of Wildlife (AAWV) guidelines are followed. If relocated, pigeons are capable of flying long distances to return to the site of capture. There is also the option of using a net launcher, which uses a blank rifle cartridge to propel a net. Fired from the shoulder much like a shotgun or rifle, net launchers can capture individual or small groups of problem birds that can be approached within about 50 feet.

Then there are decoy traps, which utilize a captured animal to attract a target species into a live trap, such as the above-described cage traps. Various designs exist for these traps, but most are fabricated from a wire exterior, supported by wood or metal frame. Entry into the trap is a one-way door, a device allowing animals to enter, but not exit. Live decoy birds of the same species that are being targeted are placed in the trap with sufficient food and water to assure their survival. Perches are configured in the trap to allow birds to roost above the ground and in a more natural position. Active decoy traps, such as these, are monitored daily, every other day, or as appropriate, to remove and euthanize excess birds and to replenish bait and water. When emptying live birds from the trap, a few are left to serve as "decoys" for future attraction. Birds removed from traps are euthanized humanely using methods listed in the section below on euthanasia. Decoy traps attract various blackbird species and pigeons. Non-targets that are trapped are released, including tri-colored blackbirds.

Both eggs and nests can be destroyed. Canada geese, mute swans, and gulls should not be allowed to nest on airport property. Provided the correct permits are in place, personnel can destroy (break eggs and remove nest material) any goose, mute swan, or gull nests with eggs found on an airport. Egg addling and destruction is the practice of destroying the embryo prior to hatching. Egg addling involves vigorously shaking an egg numerous times and causing detachment of the embryo from the egg sac, thus, making the embryo unviable. Egg destruction can be accomplished in several different ways, but the most commonly used methods are manually gathering eggs and breaking them, or oiling or spraying the eggs with a liquid that covers the entire egg and prevents the embryo from obtaining oxygen. When eggs are left intact and replaced in the nest, re-nesting by the bird is often prevented. This method can be used for nests as authorized in wildlife control permits. Eggs are addled or destroyed whenever they are encountered and positively identified as permitted species.

The nesting area is checked weekly for re-nesting until the end of the nesting season (generally the end of June). As an alternative to harassment, it may be better to shoot nesting geese and mute swans. Destroy pigeons, starlings, and house sparrows nests whenever they are encountered in airport buildings and structures. Where practical, install physical barriers, as discussed above, to prevent re-nesting. Nests of other birds hazardous to aviation generally also should be destroyed when encountered on airports. Remember that migratory bird nests are protected by federal law and may not be taken without a Depredation Permit. Each situation will have to be addressed on a case-by-case basis, depending on the species of bird and level of threat posed, location from runways, bird movement patterns, and other factors.

Currently, in the United States, only one oral toxicant, DRC-1339, also known as Starlicide (active ingredient 3-chloro-p-toluidine hydrochloride), is registered with the EPA for use in bird population management. Starlicide is a slow-acting avicide and is highly toxic to starlings, blackbirds, pigeons, crows, magpies, and ravens but only slightly toxic to non-sensitive birds, predatory birds, and mammals. Many other bird species, such as raptors, sparrows, and eagles, are classified as non-sensitive to this toxicant. The use of toxic baits to kill target birds without affecting non-target species requires considerable skill and patience. Hollow metal perches containing a wick treated with the toxicant fenthion is used to control pigeons, house sparrows, and starlings in and around buildings. Toxicants may be applied by qualified and licensed personnel in any circumstance when deemed necessary and where minimal non-target take can be guaranteed. Toxicants have varying levels of lethality for different species.

Carbon dioxide (CO2) is a common euthanasia agent because of its ease of use, safety, and ability to euthanize many animals within a short time span. The advantages of using CO_2 are: (1) the rapid depressant, analgesic, and anesthetic effects of CO_2 are well established; (2) CO_2 is readily available and can be purchased in compressed gas cylinders; (3) CO_2 is inexpensive, nonflammable, nonexplosive, and poses minimal hazard to personnel when used with properly designed equipment; and (4) CO_2 does not result in accumulation of tissue residues, thus leaving the bird edible.

Killing birds is a last resort and is only considered after habitat modification, exclusion techniques, repellent techniques and other techniques described above have failed. Shooting is normally used to remove a single offending bird. However, at times, a few birds may be shot from a flock, thereby constituting a lethal method/harassment combination, to make the remainder of the birds move away and reinforce non-lethal methods (USDA 1997). Federal, state, and possibly municipal permits are required for this approach. Shooting has been used to reduce hazards caused by birds that habitually fly over airport runways. Caution must be used so that shooting does not disturb non-target species, and shooting is not practical or desirable as a method for reducing large numbers of birds.

Now that you are a bird control expert, the next time you are near an airport, you may want to look around to see if you can spot some of these bird control designs and devices. Initial designs, continuous assessments, habitat modification, use of repellents, harassment and removal must all be used to control the bird strike hazard to aircraft. To ensure compliance, plans are written and FAA inspections are conducted. I believe that the techniques in this chapter are effective for bird control at airports; however, the techniques must be customized for each airport depending on the bird species present. While the methods described above may be effective, they are not permanent solutions. The rest of this book describes emerging technologies that could solve the bird strike problem once and for all. The existing key technologies – a ground-based bird radar, a data link to the cockpit, and an aircraft collision avoidance system – are all explained in great detail in the chapters ahead. The integration of these key technologies will make possible an electronic system to warn pilots of bird strike hazards before they happen.

Chapter 8:

Recent Advances in Bird Radar Technology

The previous chapters have described the aircraft bird strike problem, its history, and current mitigation methods used to reduce risk. This book's final chapters describe emerging technologies that, when properly integrated, could create a cockpit-based collision avoidance system for birds. This system consists of a bird radar, a data link to the cockpit, and a collision avoidance system. 'Radar' is actually an acronym for "Radio Detection and Ranging". Radio waves are used to detect the presence of a target, and one function of a radar's many functions is to determine the range to the target. Radar was invented just before WWII and was a major contributing factor to our victory. A network of early warning radars gave the United States and its allies eyes in the sky to warn our troops of approaching enemy aircraft. Command and control systems allowed the allies to respond by launching and guiding fighter aircraft to shoot them down. In this chapter, I discuss common applications, operation, and how a radar system can be used to detect birds and warn pilots of a potential collision.

The military applications for radar are classified by their mission areas – air-to-air, air-to-ground and ground-to-air. Air-to-air or air-intercept radars are used in fighters to defend the skies from enemy aircraft. This radar usually scans a given volume of airspace for potentially hostile targets. Once a target is found, the pilot can lock the radar on and decide whether to launch an air-to-air missile. An air-to-air missile is powered by a rocket motor, and an aircraft radar can guide the missile to the target. Some missiles, like the Advanced Medium-Range Air-to-Air Missile (AMRAAM), have built-in radar that allows the missile to find the target without guidance from a fighter. Because fighters are small, there is limited real estate to install a radar system, and as a result, the maximum range the radar can see is limited to tens of miles. To extend the range, another type of air-to-air radar is used by larger aircraft with bigger, more powerful radar systems. These larger aircraft, like the Airborne Warning and Control System (AWACS), do not carry armament; they provide surveillance of a bigger piece of the sky with ranges of hundreds of miles. AWACS and fighters work together with a command and control system to engage hostile targets. The AWACS sees the target first, and when the fighter is close enough to see the target, control is handed over for an intercept or shoot down.

Another military mission area for radar is air-to-ground. These radars are normally used to create images of the ground (or sea) and detect potentially hostile targets and even targets underground (ground-penetrating radar). Once a target is found, an air-to-ground weapon can be used to take out the target. Some air-to-ground weapons even have a self-contained proximity radar. The weapon is launched from the aircraft, and the radar on the bomb determines the range to the ground. At the desired range, the fuse goes live and the bomb explodes.

Yet another mission for radar in military applications is ground-to-air. These radars are on the ground and are used to find potentially hostile targets in the air. Once this is accomplished, a ground-to-air missile can be launched to shoot down an enemy aircraft. The command and control principles are similar to that of the air-to-air mission. A large ground-based early warning radar finds the aircraft at long ranges. Once the target gets closer, the target is handed off to a ground-based fire control radar that locks onto it, launches the missile, and uses its radar to guide the missile to impact.

One government (FAA) application for radar is air traffic control. These radars locate aircraft targets by using a national network of surveillance radars. In general, air traffic uses two types of radar – surveillance radars and approach and terminal control radars. The surveillance radars have a range of approximately 200 nautical miles (NM), are located around the United States, and are used to see and control aircraft during the en-route phase of flight. Approach and terminal control radars are normally associated with the airport and are commonly called 'Terminal Radar Approach Control' (TRACON).

Another government-owned and -operated radar network is the Next Generation Radar (NEXRAD) also referred to as the Weather Surveillance Radar, 1988 or WSR-88. This type of radar is designed to detect precipitation as a target. These are the radars you see on your local TV stations that display current weather conditions. Precipitation in the form of rain, snow and hail normally appear as colored areas with red signifying the most severe level. Additionally, insects, birds and bats also reflect radar signals and are sometimes called biological targets.

Radar technology can be found in many of today's civil applications as well. A police radar gun can measure the speed of a car. The gun has a radar transmitter that radiates electromagnetic waves that travel through the air at the speed of light. These waves travel away from the radar gun and strike the car as it moves through space, which changes the wave pattern. When the car is moving away, the wave is stretched in such a way that the frequency of the wave decreases. Frequency is the number of wave oscillations per unit of time. If the car is moving toward the radar gun, then the wave is compressed, and the frequency of the wave increases. This change in frequency is called the Doppler shift, named after Christian Andreas Doppler, the Austrian physicist who first discovered the phenomenon. The faster the car goes, the greater the Doppler shift. The wave is reflected off of the car, and a receiver at the radar gun which processes the signal, determines the Doppler shift, coverts

Doppler shift into speed, and places the car speed on a display for the police officer to observe.

Besides speed, another parameter that can be measured by a radar system is range. Because radar waves travel at a constant speed (i.e., the speed of light, which is 186,000 miles per second), if you measure the time it takes from transmission to reception, then the range can be calculated by multiplying the speed of light by the delay time. The speed of light is very fast, so the time delay is very short and almost imperceptible. If, for example, the time delay was one microsecond (1×10^{-6} second), then the distance would be 0.186 miles. The equation can also be used to calculate the delay time if you know the distance (distance equals the speed of light divided by time). A former Navy Admiral by the name of Grace Hopper made the nanosecond famous by carrying around a bundle of one-foot wires when she lectured about ENIAC, the first computer; at a time when no one could visualize what a nanosecond was, she used these wires to illustrate how fast electrons travel. If you do the math, electromagnetic waves travel one foot in about a nanosecond (1×10^{-9} second). Yet another parameter that can be measured is angle relative to the radar. The radar antenna creates a beam pointed in one direction. To cover 360 degrees, the antenna is either mechanically or electrically scanned. The signal processor computer keeps track of where the beam is (azimuth[18]). So, at each moment in time, a target detection has a range, Doppler (i.e., speed) and azimuth value associated with it.

While radar guns have been around for a long time, a more recent application is the use of radar as a collision avoidance system for cars. In 2008, Ford announced that a new automobile collision avoidance system called "driver assist technology" would be installed in its 2009 models. A series of radar antennas is installed around the vehicle so that the speed, range and bearing of other vehicles and objects can be processed in a computer with collision avoidance logic to control warning and display systems in the vehicle. A warning system emits an audible alarm and can automatically apply the brake

18 The azimuth is the direction the beam is steered towards. There are 360 degrees of azimuth in a circular scan of the radar. At any one moment in time, if the radar scan is frozen, then that is the azimuth it is pointing to.

to avoid a collision. The display system projects a red light onto the windshield to warn the driver. For example, if you change lanes and there is another vehicle in your blind spot, the warning system will help prevent a collision by providing a warning. Another scenario occurs when a vehicle is going too fast and a potential rear-end collision is predicted to occur. The brakes are automatically applied to avoid a crash. In this book, the proposed application for radar is as a ground-based sensor for birds. The bird speed, range and heading can be measured by the radar and the bird target reports can be data-linked to the cockpit warning and display system.

So how do these magical radar systems work? The answer is actually very complicated, and people like me spend their entire careers developing applications for radar. I will not try to make you a radar expert, but I would like to use this chapter to explain some of the more important radar parameters. Electromagnetic waves travel a certain distance and are reflected off of a target. The amount of reflected energy depends on the material, the angle that the radar beam hits the target, the angle that the radar beam is reflected off the target, and the size of the target. Material like metal has a strong reflectivity, and wood and fiberglass have weak reflectivity. The reflected energy can be characterized by a projected area called the radar cross section (RCS). The way the wave is reflected depends on the wavelength and the target size. If the wave is much larger than the target, then the reflected energy and RCS will be small. If the wave is much smaller than the target, then the reflected energy and RCS will be close to the target's physical size. If the wavelength is close to the physical size of the target, a phenomenon called resonance occurs, and the RCS will increase to its maximum level. At resonance, the RCS has a blooming and oscillatory effect which increases and decreases the value.

If the target is a bird, then the RCS will depend on the relationship of the wavelength to the size of the bird. The reflected energy will also depend on whether there is one bird or a flock of birds being observed. Additionally, a radar can determine the wing beat frequency (the rate at which the bird flaps its wings). Interestingly, the relationship between the wing beat frequency and the length of the wing can sometimes be used for bird species classification.

So why is RCS such an important parameter? Well, the larger the RCS, the greater the range a radar can see. A bird's RCS will depend on the relation between the size of the wave and the size of the bird. There is also an aspect dependency where the bird RCS will be smaller when viewed head on and larger when viewed at broadside. So in designing a radar, the target size is considered and sometimes used to choose an operating frequency.[19]

There are many other parameters and phenomena that affect the design of a radar. These parameters are combined in a radar equation that relates all the system parameters and is used in radar system top-level design and also in data analysis. These parameters include transmit power, transmit antenna gain, transmit antenna size, target radar cross-section, multi-path gain, and distance between the transmitter and the target.

Up to this point, we have assumed that radar waves travel in straight lines, but it's actually much more complicated than that because electromagnetic wave propagation considerations include a flat earth model, refraction, ionospheric propagation and attenuation. All of these methods of wave propagation must be considered when predicting a radar's performance. When the radar wave reflects off the surface, there will be constructive and destructive interference that will cause a lobe structure. When these models are used to predict the lobe structure, there will be peaks and valleys in the areas where the target can be detected. In a peak area, there will be maximum signal strength. As you move toward a valley area, the signal level falls off and eventually disappears. A flat earth model allows one to use precise angles and geometry to predict the lobe structure.[20] To simplify the calculations, we often assume that the earth is flat.

When modeling the earth, a first-order calculation is made based on a flat surface. Because the earth is round, the radar waves actually bend to follow the curvature. This bending of a wave is called refraction. When a wave bends, the actual distance it travels is actually greater than a straight line. This difference in distance must be considered

19 Frequency equals the speed of light divided by the wavelength.

20 A lobe structure is the antenna pattern that is formed.

when predicting a radar's performance. Another consideration for wave propagation is the ionosphere, which contains particles that are ionized by the sun. This region of the earth's atmosphere will cause certain radar wavelengths to reflect back or bounce off. A radar wave with a frequency of less than 50 MHz will be reflected off the ionosphere. Frequencies higher than 50 MHz will pass through the ionosphere and not be reflected. The reason this is important to radar designers is that if you design a radar with a frequency of less than 50 MHz, then you will be able to extend the range beyond or over the horizon. At the lower frequencies, the radar wave bounces off the ionosphere and returns to the ground at extended ranges. This bounce can occur several times before the radar wave hits a target and is reflected back. After the radar wave is reflected, it is reflected back up to the ionosphere and down to the radar receiver. Maximum ranges of thousands of miles can be attained by using this principle. In the 1950s, the United States designed a series of radars on the east and west coasts called Over the Horizon Radar Backscatter (OTH-B). The theory was that the OTH-B radar would be able to detect large enemy bombers over the North Pole that could invade the United States The long-range detection would provide an early warning of attack so that the United States could launch an interceptor to perform a shoot down.

A radar is basically a communication system. It has a transmitter, receiver, signal processor computer and a display system. The radar transmitter generates a signal that is amplified and transmitted by an antenna. Antenna design is a very complex topic, but the basic idea is to form a focused beam. The shape of the beam is controlled by the design, and normally, the transmit antenna is also used for receiving the reflected wave from the target. Radar antennas come in all sizes and shapes. In order to allow the radar antenna to cover 360 degrees, the beam must be mechanically or electronically scanned. Mechanical scan is accomplished by placing the antenna on a platform with a motor that moves the antenna in a circular pattern; however, the disadvantage to this implementation is that mechanical parts wear over time and must be replaced. Electronic scan is accomplished by using a phased array, which scans by changing the phase at each antenna array element.

The current commercial bird radar antennas come from boat radars. The antenna is called a T-bar and uses a mechanical scan to cover 360 degrees in azimuth. The beamwidth of these antennas is 20 degrees. The current implementation is to use one T-bar for horizontal scanning and another for vertical scanning. The horizontally scanning antenna covers 360 degrees. To get a partial three-dimensional picture (i.e., range, azimuth and elevation), the bird radar companies use a second T-bar antenna that is perpendicular to the first antenna and scans vertically. Three-dimensional coverage will occur in the area where the horizontal and vertical beams overlap. Although each radar has a separate display, which is really two dimensional coverage. The result is three-dimensional coverage over a very narrow area that is normally aligned with the runway. To implement this antenna pattern in an airport environment, the vertically scanning antenna must be aligned with the runway. The problem is that in order to obtain complete coverage at an airport with multiple runways, multiple vertically scanning radars (i.e., one for each runway) must be used. Multiple antennas will significantly increase the cost of implementation. What is needed is a single antenna that provides three-dimensional coverage over 360 degrees.

After a radar wave is reflected back, in addition to containing information about the target, the return signal contains unwanted elements called 'clutter' and 'noise'. These elements are normally removed by a computer and complex software signal processing algorithms. Surface clutter results from the radar return from the ground or sea. The reflected radar signal from land clutter depends on the type of terrain. Different types of terrain, such as farm land, mountains and forests, will all have different reflection and absorption characteristics. Experimental measurements of the different types of clutter have allowed radar engineers to build a table of radar cross-section per-unit-area values. In addition to terrain type, the amount of energy reflected also depends on the grazing angle and radar frequency. Over the sea, the reflections from the ocean will depend on the height of the waves and the frequency of the radar.

In addition to surface clutter, the radar also receives reflections from objects on the ground, like towers and buildings. If this type of clutter is too strong, then radar sensitivity could degrade, which will decrease the radar's maximum range. Generally, insects, birds and bats are also treated as unwanted clutter and are removed by using signal processing. There is an old saying that goes "one radar's clutter is another radar's target." Normally, birds are treated as clutter, but in the case of a bird radar, the bird is actually the target. As long as radars have been in existence, birds have been treated as clutter and filtered out. The FAA radars that are used today, for example, filter out unwanted bird targets. These radars are 10 to 40 years old, and many will be decommissioned with the coming of the Next Generation Air Transportation System in order to save money on operations and maintenance. Another illustration of this old saying occurs with the implementation of a weather radar. When searching for airborne or ground targets, weather is normally treated as a type of clutter and suppressed; however, if you are designing a weather radar, precipitation in the form of rain, hail and snow becomes the target.

Perhaps you now have a better appreciation of the complexity of the design of a radar system. Clutter is just one of many parameters that must be understood and modeled to predict performance. All of these parameters are used in the radar equation described above to generate a top-level analysis for design. So how can clutter be removed from a composite signal that contains both clutter and the target? The answer is actually pretty simple for ground-based radar. An airborne target is moving, and the ground is not moving. A signal processing algorithm can be designed to discriminate between the target and the clutter based on movement and lack of movement (i.e., Doppler shift). A signal processing algorithm can be designed to remove the clutter so that only the target remains. For airborne radar, though, the clutter problem is much more complicated. The radar is moving, the clutter is moving, and the target is moving. Complicated signal processing algorithms are normally used to remove the moving clutter.

Another unwanted component of the composite signal that must be removed is called 'noise'. There are basically two types of noise: atmospheric and receiver noise. Atmospheric noise has several sources: the galaxy, the sun, the atmosphere and man-made noise. Galactic noise or cosmic noise comes from the Milky Way. The higher the frequency of the galactic waves, the lower the galactic noise. The sun is also a source of noise. As the frequency of a radar increases, the sun's noise increases as well; and when sunspots occur, there is a significant increase in this type of noise. The atmosphere actually acts as an absorber of noise, and the amount of absorption is also frequency-dependent. The last type of noise is man-made noise, and it occurs from any man-made electrical source like a car or lawn mower. A large amount of work is done in radar system design to keep the noise level down.

The last type of noise that must be processed out of the signal is receiver noise, also called 'thermal noise'. Thermal noise is generated by the movement of electrons (Brownian motion) in the radar receiver. It is called 'thermal noise' because the greater the temperature, the greater the movement of electrons and, thus, the greater the thermal noise. So, cooling a radar receiver will lower the temperature and decrease the thermal noise. All of these sources of noise can be modeled statistically and must be taken into consideration when making design choices. They must also be taken into consideration when designing signal processing algorithms to remove the clutter and noise from the composite signal.

Once all of the radar parameters are understood, then the design process begins. Many subsystems are connected together to complete a radar system. The first subsystem is called the radar receiver. Radar receiver design is very complicated. The receiver must process a very weak signal and amplify, demodulate (i.e., remove the high-frequency components) and digitize it (i.e., convert from analog to digital). The next subsystem is called the 'signal processor'. This subsystem is a computer with software (i.e., signal processing algorithms) that removes the clutter and noise. After the clutter and noise are cancelled, the radar must then determine if a target is present. Normally, a signal threshold is statistically calculated, and a threshold device is used to determine whether a target is present or absent. The performance of this device can be characterized by the probability of detection and the probability of false alarm. The probability of false alarm is the likelihood that a target will be declared when only noise is present. As the detection threshold is decreased, the probability of false alarm increases. Obviously, you don't want to have false target declarations, so you raise the threshold to reduce the false alarm rate, but now there is a tradeoff to be made. As you increase the threshold, you also decrease the likelihood of declaring that a target is present; that is, you lower the probability of detection. During radar development, it is very tricky to get these two probabilities where you want them. Radar performance can vary significantly in different clutter and noise environments. So radar design becomes both an art and a science.

Now that you are better acquainted with radar, you can understand the importance of the Congressional hearings on Flight 1549 in February 2009 that discussed the use of bird radar technology. After a series of bird strike accidents in 1999 (over 10 years ago), the NTSB recommended that the FAA pursue technologies to fix the bird strike problem. In response, the FAA started a research and development project in 2000 to attempt to develop a bird radar capability using existing FAA radars. Investigations into how, and how much it would take, to modify the FAA radars to see birds have been undertaken with a prohibitively expensive result. After 7 years of R&D, there was still no solution. So the FAA changed the project from R&D to evaluation of commercial bird radars. A collaboration of government, academia, industry and the airport community was established in 2006 to address the problem of civil aircraft bird strikes. Participants include the FAA's Technical Center in Atlantic City, NJ; the University of Illinois' Center of Excellence for Airport Technology (CEAT) at Urbana-Champaign; a commercial bird radar company; and Seattle's Sea-Tac Airport. CEAT is one of several FAA-supported Centers of Excellence, each of which has specific objectives and expertise. The objective of this project is to find a low-cost radar that can detect birds as far as 3 to 5 nautical miles from airports and develop a process to advise pilots of potential impending bird strikes. The five-nautical-mile distance is important to air traffic controllers because it defines the limit of the lower tier of controlled airspace. The first commercial experimental bird radar was installed at Seattle-Tacoma Airport and Whidbey Island in 2007. The same experimental radar was installed at Chicago O'Hare airport and John F. Kennedy Airport in New York. The current direction that the FAA is taking is to continue to assess bird radars to improve wildlife management programs and safety at airports. As described above, the three-dimensional coverage of this experimental radar is very limited. Three-dimensional coverage over 360 degrees is necessary to obtain the position and altitude of the bird target. Pilots will need to know where and at what altitude the birds are in order to avoid them.

Throughout all of this, no one has discussed the operational implementation or use of bird radars to improve air safety. These radars are simply being evaluated so that airports can purchase them. There is no plan for how they will be used, so I will describe what I see as a logical progression for bird radar usage at airports. I briefed one major airport on a bird radar that I am currently working on and was told by the airport management that their operational implementation concept was to have wildlife control technicians monitor the birds. If a potential collision is observed, the wildlife technician calls the control tower so the controller can warn the pilot. I believe that bird radars will be an excellent tool to improve the wildlife management program at airports. However, the final implementation should be to simultaneously send the bird targets to both the control tower and the pilot. Work on this implementation is my operational concept. I plan to dedicate a great deal of my time to education and implementation of using bird radars to send data links to cockpits for collision avoidance.

For almost 10 years, the FAA has been studying radar technology to detect and track birds. In the meantime, bird populations are multiplying. The FAA is collaborating with the USDA Wildlife Services team to determine operational suitability of using bird radar technology at airports. 'Operational suitability' means how much data needs to be transmitted and what pilots and controllers should do if a potential bird strike scenario occurs. The media, politicians and ALPA are placing increased pressure on the FAA to implement a solution to this problem. The FAA, however, is pushing back and asking for more time to assess operational use, yet they have neither pilots nor controllers on their assessment team. The commercial bird radar companies are trying to sell their radars to airports by announcing in press releases and interviews that their radars are ready for operational implementation. They are leveraging their sales with the military and using the airliner in the Hudson accident to gain attention and market their systems. To this day, though, no US airport has a bird radar installed in the control tower.

During the NTSB hearings in June 2009, the FAA testified that they wanted the following tasks to be part of the evaluation process: understanding of expected radar performance; validation of target quality; understanding of the impact of the local environment on system performance; assessment of electromagnetic compatibility with other airport systems; assessment of data management, integration and interoperability; working with wildlife management personnel for proper implementation; establishment of minimum requirements; determination of updates to regulations; and development of guidance on how to acquire, deploy, integrate, acceptance test, operate and maintain bird radars. I believe that it is alright to go down the current path of evaluating bird radars so that they can be approved for use as a tool to improve wildlife management programs at airports; however, in parallel, the government should invest in the long-term development of these radars, data links to the cockpit and useable warning and display system for pilots and controllers. New ground is being broken in this area every day. Both a long- and short-term plan needs to be established. At present time, though, no plan exists. The FAA and the airports know that they need a plan, but they haven't taken the necessary steps to establish one and they don't really understand the technology. They are getting smarter, and everyone has a different opinion about how these bird radars should be operationally employed.

Currently, bird radars are seen as an improvement to wildlife control at airports. A bird radar seems to be a magical solution because it gives ground personnel the eyes to see the birds. When the bird target tracks are displayed with an airport overlay, there is immediate situational awareness as to whether the bird flight path will conflict with an aircraft departure or arrival. This is a judgment call, and judgment is a learned skill. It takes thousands of hours for a pilot to become very skilled. In the same manner, an airport wildlife expert will need extensive training to warn the controllers of an impending collision. If bird radars are used in the control tower by integrating bird targets with air traffic targets on the same display, an exhaustive safety certification process will need to occur with the FAA. High radar false alarm rates will not be tolerated. Human factors studies will need to be accomplished to create an optimal human interface and display.

Controllers will need to be trained on the correct reaction to an impending collision between an aircraft and a flock of birds. Controllers will need to develop new concise terminology to use on the radio when warning a pilot of an impending collision. The FAA will need to write new, and rewrite old, regulations and procedures to assign responsibilities. When all this occurs and a bird strike happens, who will be blamed for the cause of an accident? Will the cause be a bird radar malfunction, a wildlife hazard technician error, or a controller error? Controllers are currently responsible for airspace around busy airports out to 5 NM; what will be the limits of a controller's responsibility – one mile, three miles or five miles from the control tower? I believe we need to put a technology program together to simultaneously send the bird targets to the end users – the pilots and controllers. The technology exists today as described in subsequent chapters. The bird targets need to be displayed in the control tower so that controllers can warn pilots. Controller warnings may work at low-traffic-density airports, but at the busiest airports, real-time warnings are impractical since controllers have no time to monitor a separate bird display for potential collisions.

I believe that the solution with the most promise is to electronically send the targets to the cockpit so that the pilot can make the final decision. Each time you add another decision layer to the communication chain, there will be a time delay, and time is life when you are flying at over 200 mph. If bird targets are sent to both pilots and controllers, new operational procedures will need to be developed to handle potential conflicts. If a conflict occurs on takeoff, what is the procedure? If the bird targets are observed from the cockpit during takeoff, the pilot can make a decision to delay the takeoff in order to avoid a collision, but if the conflict occurs on arrival, what is the procedure? If the aircraft is approaching to land, the pilot can do a turn in holding, go around or make a turn to avoid a potential conflict. What if the potential conflict occurs in climb, descent or cruise flight? During climbs, descents and cruise flight, an airline pilot flies his aircraft by looking at instrumentation that shows his present position and future flight path. It is only necessary to send the bird targets to the aircraft and overlay them on this display. Then a pilot can see if his future flight path will intersect with the path of a flock of birds. If that scenario is observed in flight, then a pilot can alter the flight path to avoid a collision. Will the pilot need to get permission from the controller to alter his flight path to avoid a collision, or will he be allowed to change course on his own? Will he be allowed to climb, descend or turn?

As you can see, there are many questions that deserve thorough consideration. What if the optimal solution of putting bird targets in the cockpit occurs? A collision avoidance system for birds will have to go through several levels of safety certifications before bird targets can be sent to the cockpit. As previously mentioned, human factors experts will need to study the optimal way to display the targets. Avionics manufacturers will need to add a new capability to receive and display the targets. Collision avoidance systems will require new software and symbology to provide both audible and visual cockpit warnings. Airline procedures will need to be written for a pilot to respond properly to an impending collision. Pilots will need to be trained in the simulator on an annual basis to ensure proficiency in reacting to an impending bird collision.

Will NTSB accident investigations cite pilot error as a cause for an aircraft bird collision? What role will controllers play if the bird targets go directly to the cockpit? Should pilots tell controllers that they had a potential collision so that other pilots can be warned or so that controllers can delay departures or send arrivals around? Should the bird targets be sent to both the pilot and controller simultaneously?

Present aircraft collision avoidance systems only allow aircraft to climb or descend in order to avoid a collision. The software logic commands one aircraft to climb and the other aircraft to descend. There are instrument indications to display whether the pilot is climbing or descending quickly enough. There are audible warnings that announce commands like "climb, climb." The aircraft is not allowed to turn unless the conflict is visually identified because the collision avoidance system is not precise enough. The displayed aircraft target symbol changes color and symbology depending on how close the conflict is. The pilot must be trained on how to fly the collision avoidance maneuver. Presently, the pilot cannot use the autopilot to fly the collision avoidance maneuver; he must disengage the autopilot and auto-throttles and fly the maneuver manually; however, there is ongoing research into using the autopilot to automatically fly the avoidance maneuver, and this concept has actually been demonstrated as being a safer approach. Having the computer fly the aircraft prevents an overshoot or undershoot, which improves safety. Each pilot has his own technique for flying this maneuver and can be either too aggressive, or not aggressive enough, to escape a conflict.

In conclusion, a good bird radar may reduce risk, but it will not solve the problem alone. Commercial bird radar companies label their radars as bird strike prevention radars or bird strike avoidance radars. I believe that the only system that can prevent a collision between an aircraft and a bird is the pilot. A radar alone cannot do this. The bird targets should be sent directly to the cockpit (and simultaneously to the controller). The pilot must be automatically warned of an impending collision so that he can take the proper action. As an airline and former military fighter pilot with over 10,000 hours, I believe that this is the optimal solution. This plan is not currently being considered. We need to commit our scarce resources to solve the aircraft bird strike problem, and the time to act is now.

Chapter 9:

BAM, BASH, AHAS and Advanced Concepts

Thus far, we have discussed civil bird strike prevention in the vicinity of an airport; however, military aircraft have also had their share of bird strikes. The USAF alone has had over $16 million in damage from bird strikes, just in 2007. The turkey vulture is the species that has caused the most damage with over 800 bird strikes to USAF aircraft, resulting in over $50 million in damage. Over 25% of these accidents have occurred during low-level operations when the aircraft are flying at speeds in excess of 400 mph. In the mid-1980s, the USAF Bird Aircraft Strike Hazard (BASH) team and the FAA began the effort to reduce the incidence of off-airfield bird strikes with the development of a Bird Avoidance Model (BAM). The BAM is based on over 30 years of historical data and includes over 60 species of birds. The model is used by aircrews before a flight to determine times and locations of elevated bird activity. The BAM also overlays bird activity levels onto low-level military training routes.

In the early days, the BAM was not very good. Bird migration corridors were overlaid onto navigation charts to depict areas of concentration. Data was collected manually and was more difficult to analyze. Each type of flight route described above had to be analyzed separately. After many years of improvement, the current BAM is now Web-based and used by many operational military squadrons. Hazard levels in the BAM are based on bird mass. Larger birds mean a greater hazard. The average mass of the birds per square km is transformed into 9 levels of bird strike risk. The BAM provides both short-range risk forecasts and historical risk levels. The analysis and display of hazardous areas can be done in minutes and is user-friendly. The BAM uses current information from many sources and is updated continually as new information becomes available. It uses a 30-year historical database for its predictions. While this may be a useful planning tool, the information is not real-time; and bird populations and migration routes change over time, so its usefulness is limited.

The NEXRAD radars described in Chapter 5 are normally used to detect and display weather. Advanced signal processing algorithms have been developed to remove weather and display biological targets (e.g., birds, bats and insects) from the radar data. These algorithms and display software have been integrated into what is called the Avian Hazard Advisory System (AHAS), which is like a real-time version of the BAM. It is used by the military to predict risk levels for low-level and bombing-range operations. AHAS is not designed for use at an airfield; however, aircraft flying in a military base traffic pattern are normally 10 to 20 miles away at 2000 to 3000 feet, so this data may be useful for display in the cockpit.

Wildlife control at military airfields is handled by the BASH team, which consists of the equivalent of wildlife control specialists at civilian airports. The BASH program for the USAF is administered out of the Air Force Safety Center at Kirtland AFB, NM and applies to all regular USAF and USAF Air National Guard and Reserve personnel. Air Force Pamphlet 91-212 covers the USAF BASH program, and the following paragraphs are excerpts:

Reduction of strike hazards may be divided into four categories: Awareness, Control, Avoidance and Aircraft Design. Wildlife strike hazards to aircrew and aircraft may be significantly reduced using a combination of the methods listed above. This may result in substantial savings of Air Force resources. The Bird/Wildlife Aircraft Strike Hazard (BASH) Team was formed to coordinate efforts in all areas. The BASH Team assists Air Force organizations worldwide to reduce the risk of bird strikes and collisions with other animals, such as deer. Wildlife strikes can never be eliminated; but an aggressive, well-planned program developed on the basis of wildlife habits, the environment, and the base mission may limit the potential for these strikes to occur.

A well-written, workable BASH plan is the key to reducing strike hazards and ensures continuity of knowledge with personnel turnover. Plans should be tailored to meet the specific hazards encountered locally. Integrate all habitat modification procedures making sure natural resource plans are compatible with base BASH program plans. As a minimum, this plan should: inform new personnel of local hazards; identify local conditions on the airfield attractive to wildlife and cite measures to reduce these attractions (e.g., long grass, insect reduction, water drainage); outline bird dispersal procedures; and specify base bird watch condition codes, location where daily code will be displayed, implementation procedures, authorization for declaring codes and flight operations under specified bird watch condition codes.

The Bird Hazard Working Group (BHWG) normally consists of representatives from flight safety, airfield management, base operations, air traffic control, civil engineering, aircraft maintenance, public affairs, and base legal. The base Natural Resource Manager and base Wildlife Biologist should also attend. Other representatives, such as the golf course manager, munitions or civilian operations personnel, may be included based on BHWG issues. The group's purpose is to assist the safety office in drafting and implementing the Bird/Wildlife Hazard Reduction Plan.

BHWG meeting topics should include but are not limited to: USAF-reported wildlife mishaps and incidents; USAF BASH Team information updates; locally observed/reported wildlife activity, to include LL ranges/routes, airfield inspections/surveys, recovered wildlife remains, wildlife strikes; local wildlife habitat management/modifications, to include dispersal/depredation activity, environmental/land management activity, land uses (landfills, agriculture crop seasons); BASH-related budgeting issues; annual bird migrations; local BASH plan procedures and responsibilities, to include observed effectiveness/deficiencies; BASH awareness training/education, to include BWC code definitions and communications; flying schedule and wildlife activity conflicts and the BASH self-inspection checklist.

Evaluating an airfield for wildlife hazards involves the identification of local wildlife activity periods, species causing hazards, locations of favorable habitat, and documenting this information into daily activity surveys. When information is compiled over several seasons or years, to include night and day surveys, development of a more effective wildlife hazard reduction program is possible. The wildlife surveys should include date and time; weather conditions; wildlife species of concern; favorable habitat locations on the airfield; wildlife movement corridors (soaring to and from roosts, feeding, etc.); wildlife activities (loafing, feeding, drinking, etc.); potential attractions; wildlife responses to mowing; and/or land uses adjacent to base property (planting/plowing etc.).

Documenting the local wildlife problem, technical assistance received, and the successes of solutions tried are essential parts of any wildlife hazard reduction program. Complete documentation is necessary to acquaint new personnel with the problem and may be required in any civil litigation regarding the resolution of a wildlife hazard. Documenting the use of non-lethal techniques may also be necessary to obtain permits to take lethal action, if needed, to resolve an extreme hazard. Photograph and summarize all hazardous situations wildlife create on base. For example, pictures of gulls loafing on the airfield accompanied by observations showing the birds using a nearby sanitary landfill can provide a strong case against future expansion of the landfill. Good documentation gives credence to the problem and shows solutions are being considered.

The USAF uses the same civilian control methods described in Chapter 7, but it is interesting to note that they call out the ineffective methods of control. Stuffed owls and rubber snakes have been advertised to rid hangars and buildings of birds. They are usually a waste of money and effort. Rotating lights have brought conflicting results but are generally considered ineffective. Birds quickly habituate to these devices, and the problem remains unsolved. Eyespots on aircraft components are being studied in the United States and other countries. Early results suggest the addition of eyespots does not significantly reduce the BASH potential. Ultrasonic devices have thus far proven unsuccessful in deterring wildlife from colliding with aircraft. Very few bird species can hear ultrasonic sound. Of the bird species most often struck on airfields, most cannot detect the presence of ultrasonic sound, therefore rendering these devices useless for dispersing birds on an airfield.

Another interesting point is specific to military flying. On takeoff and departure, aircraft making formation departures increase risk of damaging bird strikes when birds are feeding or loafing on or near the runway. Formation and single-ship interval takeoffs often result in birds being scared up by the lead aircraft, causing the wingman to hit the birds. When the lead aircraft scares up large flocks of birds, the wingman should delay departure until the birds are clear of the runway. Pilots of lead aircraft must be alert to warn the wingman of bird hazards during takeoff. This is especially important as the wingman's attention is focused on the lead aircraft. When flocks of migratory birds are a problem, the risk of serious bird strikes is increased by formation takeoffs and single-ship interval takeoffs with minimum spacing involving rejoins, increase the risk of serious bird strikes. All rejoins require greater attention by pilots to the lead aircraft's position. The increased speed required to catch the lead aircraft after takeoff increases the risk of damaging bird strikes. When birds are known to be flying in the area, departures under visual meteorological conditions (VMC) may be modified to reduce the risks. Departures should be made in trail, with the rejoin beginning after the aircraft passes 2000 to 3000 feet above ground level (AGL). If aircraft are to enter a low-level route immediately or stay at an intermediate altitude for a prolonged period, tactical formation provides enough aircraft clearance to allow wingmen to stay clear of birds.

During the en-route phase of flight, the USAF pamphlet states that aircrews experiencing en-route bird strikes should abort the mission when practicable. While some engine ingestions or a windscreen strike may be readily apparent from the flight deck, the damage from many engine, fuselage, wing, tail, or radome strikes cannot be adequately assessed. Continuing a mission may cause greater structural damage and lead to a serious in-flight emergency later.

Also, during low-level operations or when operating in special-use airspace, higher aircraft speeds and greater exposure within bird flight environments lead to many damaging bird strikes. During these flights, aircrews are involved in cockpit duties, allowing little time to monitor bird activity. "Heads-up" flying should be stressed as much as possible during these critical operations.

A final section of the USAF pamphlet covers military preparation for flight. It states that wildlife hazards should be considered and incorporated into the mission planning process. This should include applicable bird advisories and hazard information, available through Internet sources, Automated Terminal Information System (ATIS), or disseminated locally. Briefings on bird strikes and other wildlife hazards are much like briefings on takeoff emergencies when urgency dictates a pre-planned course of action. As a minimum, aircrew briefings should include the following: potential bird hazards along their proposed route of flight; use of the double helmet visors or sunglasses during daylight hours, the clear visor at night or during low-level operations; avoidance maneuvers at low altitude; actions if flocks of birds are encountered (for example, initiate a climb since the majority of birds dive to avoid a potential collision); and mission abort due to bird/wildlife strike. The aircrew's ability to react to a bird strike situation is further enhanced by periodically reviewing bird strike procedures during continuation training and safety briefings.

Have you ever been on an airliner and encountered turbulence? Well, turbulence is a $150-million-per-year problem. People are hurt, and aircraft are damaged. You can't see turbulence, so how can you avoid it? The rest of this chapter discusses how the FAA defines the turbulence problem, how it is being dealt with and some recent technical developments that actually solve this seemingly unsolvable problem by using existing aircraft technologies and data links.

Obviously, solving the turbulence problem does not solve the bird strike problem, but it is presented here to promote the concept of using existing technologies and data links to solve the bird strike problem. Bird target data could be data-linked directly to the cockpit using a subset of the turbulence solution architecture presented below. FAA Advisory Circular 120-88A published in 2006 was issued to help airlines prevent injuries caused by turbulence. The following is an excerpt from that AC:

> The NTSB defines an accident as an occurrence associated with the operation of an aircraft which takes place between the time any person boards the aircraft with the intention of flight and all such persons have disembarked, and in which any person suffers death or serious injury, or in which the aircraft receives substantial damage. The NTSB defines a fatal injury as any injury that results in death within 30 days of the accident. The NTSB defines serious injury as any injury that (1) requires the individual to be hospitalized for more than 48 hours, commencing within 7 days from the date the injury was received; (2) results in a fracture of any bone (except simple fractures of fingers, toes, or nose); (3) causes severe hemorrhages, nerve, muscle, or tendon damage; (4) involves any internal organ; or (5) involves second- or third-degree burns, or any burns affecting more than 5% of the body surface.
>
> 'Light chop' is defined as slight, rapid, and somewhat rhythmic bumpiness without appreciable changes in altitude or attitude. 'Light turbulence' is defined as slight, erratic changes in altitude and/or attitude. Occupants may feel a slight strain against seatbelts. Unsecured objects may be displaced slightly. Food service may be conducted, and little to no difficulty is encountered in walking. 'Moderate chop' is defined as rapid bumps or jolts without appreciable changes in aircraft altitude or attitude. 'Moderate turbulence' is defined as changes in altitude and/or attitude, but the aircraft remains in positive control at all times. It usually causes variations in indicated airspeed. Occupants feel definite strain against

seatbelts. Unsecured objects are dislodged. Food service and walking are difficult. 'Severe turbulence' is defined as large, abrupt changes in altitude and/or attitude. It usually causes large variations in indicated airspeed. Aircraft may be momentarily out of control. Occupants are forced violently against seatbelts. Unsecured objects are tossed about. Food service and walking are impossible. Extreme turbulence occurs when the aircraft is violently tossed about and is practically impossible to control. This may cause structural damage.

The types of turbulence include: thunderstorm turbulence which is associated within and in the vicinity of thunderstorms or cumulonimbus clouds. A cumulonimbus cloud with hanging protuberances is usually indicative of severe turbulence; clear air turbulence; high-level turbulence (above 15000') not normally associated with cumuliform cloudiness; mountain wave turbulence is a result of air being blown over a mountain range or a sharp bluff causing a series of updrafts and downdrafts.

Flight attendant injuries occur at a disproportionately high rate compared to other crew members and other cabin occupants because flight attendants spend more time in the passenger cabin unseated and, therefore, unbelted. Effective training emphasizes to flight attendants that the overlying objective throughout all flight attendant training is to ensure that flight attendants are confident, competent, and in control while conducting their activities in the cabin. However, during a turbulence encounter, the most appropriate first response by a flight attendant might be self-preservation. Training can include the effective use of the passenger address system (PA) and other methods of communicating with passengers; the location of handholds throughout the airplane (or equipment that could be used as a handhold); and how to secure a service cart or an entire galley in minimum time. Flight attendants can be made aware of ways in which human psychology might play into a turbulence encounter, and might actually increase their risk of injury. For example, on a short flight, with little time to complete a cabin service, flight attendants might be less conservative regarding their personal safety than on a longer flight with no time constraints. Communication and coordination among crewmembers is a critical component of an effective response to turbulence or a threat of turbulence. Air carriers can develop and implement CRM training in Initial and Recurrent crewmember training that encourages a coordinated crew response before, during and after a turbulence encounter.

There are operating procedures that the airlines can use to prevent injuries caused by turbulence. Procedures promoting voluntary seatbelt use and compliance with the fasten seatbelt sign can include the flight crew promptly and clearly communicating turbulence advisories to flight attendants and passengers. Those advisories can include directions to be seated with seatbelts fastened and to secure cabin service equipment. Flight attendants effectively communicate directions to passengers to be seated with seatbelts fastened. Air carriers develop and implement practices to encourage the use of an approved child restraint system (CRS) to secure an infant or a small child that is appropriate for that child's size and weight. Parents and guardians can be encouraged to ensure that children under the age of two, traveling with approved CRS, occupy the CRS any time the fasten seatbelt sign is illuminated. Flight attendants can verify that the CRS is secured properly in a forward-facing seat and that the child appears to be properly secured in the CRS.

Air carriers can develop and implement practices to improve passenger compliance with seating and seatbelt instructions from crewmembers, such as video presentations incorporated as part of a flight attendant's safety demonstration that can illustrate the benefits of using effective turbulence practices; articles in airline publications, pamphlets in seatback pockets or information on safety information cards can encourage passengers to engage in effective practices, such as keeping seatbelts fastened at all times; before descent, or early in the descent, depending on conditions, flight crews can give passengers notice by way of an announcement that the fasten seatbelt sign will be illuminated in 10-15 minutes, and that any personal needs requiring movement in the cabin should be met before that time. This practice emphasizes the requirement to comply with the fasten seatbelt sign; air carriers can implement spoken and written advice to passengers that FAA regulations require them as individuals to comply with crewmember instructions regarding the fasten seatbelt sign; air carriers can promote reasonable communication between flight attendants and the flight crew regarding the use of the fasten seatbelt sign; the environment in the cabin may be very different from the environment in the flight deck during turbulence. Flight attendants should feel free to request that the flight crew illuminate the fasten seatbelt sign whenever it is appropriate to do so in their judgment.

Volunteers representing various stakeholder groups within an air carrier may work together as a highly competent team. The team can review the air carrier's own turbulence encounters and resulting injuries. That review might shed light on root causes of the encounters and the injuries, and, in turn, might show the way to effectively prevent them. Current information is generated in a variety of ways such as turbulence reports from crewmembers, injury reports from flight attendants, post-encounter interviews and other processes that provide information for review and analysis. Information useful for analysis can include length of flight, route of flight, time of year, phase of flight, aircraft type, type of injuries received by passengers, type of injuries received by crewmembers, adequacy of crewmember communications and adequacy of air carrier procedures.

Dispatchers can communicate with flight crews, and flight crews can communicate with dispatchers, before, during, and after a flight, and can be encouraged to do so whenever necessary. In the preflight planning phase, the dispatcher may use the "remarks" section of the dispatch (flight) release to advise flight crews of known or forecast turbulence. A "call dispatch" notation on the dispatch release may be included to indicate that the dispatcher believes a telephone conversation with the pilot is necessary. Communication may resume at any time during or after flight using an Aircraft Communication Addressing and Recording System (ACARS), company radio, or telephone – and should be encouraged by an air carrier's management to improve the flow of real-time information regarding turbulence.

Preflight weather briefings, verbal or written, must include forecasts of turbulence and pilot reports of turbulence caused by thunderstorm activity, mountain wave activity, clear air turbulence, low altitude frontal windshear and low altitude convective windshear. During a flight, the pilot and dispatcher must communicate any changes in the forecast or actual turbulence conditions via voice or digital communication methods in order to pass real-time turbulence information along to other flights.

Continual reinforcement of the air carrier's turbulence avoidance policy can be evident in theory, in on-the-job training, and in practice. It is important to assure that the pilot weather briefing includes known areas of turbulence. The flight routing should be discussed, including en-route altitudes, with flight crew prior to departure. Dispatchers should plan flights so they will not proceed through areas in which there are thunderstorms of more than moderate intensity. Dispatchers should plan flights to avoid areas of severe turbulence and severe thunderstorms.

The first and most fundamental step in developing effective practices is that an air carrier can adopt a corporate culture of avoidance of turbulence as the first line of defense. Implementing a turbulence avoidance culture can include standard operating procedures (SOPs) for dispatch and flight operations providing for rerouting around forecast and observed turbulence, and for observing standard clearances between thunderstorms and aircraft. In the past, the practice of rerouting has met with limited air carrier acceptance, primarily because of the inaccuracy of first-generation turbulence forecast products, the subjectivity inherent in Pilot Weather Reports (PIREP) (if available), and the operational costs of rerouting. However, recent advances in automation, atmospheric modeling, and data display have improved forecast accuracy, data delivery, and PIREP subjectivity, improving the odds that a well-chosen rerouting would in fact avoid turbulence.

If flight into forecast turbulence is unavoidable, timely notification to the cabin crew is crucial to their safety. If turbulence is expected before the flight departs, the preflight briefing to the lead flight attendant must include turbulence considerations. The briefing can be the same as an in-flight briefing for expected turbulence, including actions the captain wants the cabin crew to undertake any time turbulence is expected; intensity of turbulence expected; methodology for communicating to the cabin the onset or worsening of turbulence (e.g., cabin interphone or PA); phraseology for the cabin crew to communicate the severity of turbulence; expected duration of the turbulence and how an "all-clear" will be communicated; utilize a positive signal for when cabin crew may commence their duties after takeoff and when they should be seated and secured prior to landing; passengers will be informed of routine turbulence via the PA system. Do not rely on the seatbelt sign alone. Cabin crew will be informed of routine turbulence via the interphone. If at any time the cabin crew experiences uncomfortable turbulence without notice from the flight crew, they must immediately take their seats and inform the flight crew. All service items must be properly stowed and secured when not in use.

If turbulence is expected, the Captain can thoroughly brief the cabin crew on the expected turbulence level and its duration. The cabin crew should be instructed to immediately and plainly report any deviations from the expected turbulence level. A method should be developed to inform the cabin crew of the completion of the turbulence event. When there is little warning, the Captain turns on seatbelt sign and makes a public address announcement, "Flight Attendants stow your service items and take your seats. Passengers please remain seated until this area of turbulence has passed and I have cleared you to move about the cabin." Cabin crewmembers stow all applicable service items, perform cabin compliance check, and secure themselves in their jump seats. Lead flight attendant informs the Captain of the completion of these items. When conditions improve, the Captain uses the public address system to advise the cabin crew that they may resume their duties and whether or not the passengers may move about the cabin. When the turbulence is imminent, the Captain turns on the seatbelt sign and makes a public address announcement: "Flight attendants and passengers, be seated immediately. Passengers please remain seated until this area of turbulence has passed and I have cleared you to move about the cabin." Cabin crewmembers take the first available seat and secure themselves.

No compliance checks are performed and items are secured only if they present no delay in securing a person in a seat. When conditions improve, the Captain makes a public address announcement advising the cabin crew that they may resume their duties and whether or not the passengers may move about the cabin.

Continued improvement in turbulence-related weather products requires better handling of real-time information on the state of the atmosphere at any given time. The most promising way to capture and convey this information is through a comprehensive program of reports from aircraft in flight. That program would be founded on automated turbulence reporting supplemented by human reports (PIREPs). Air carriers can promote real-time information handling by the following steps: commit to the installation of the Turbulence Auto-PIREP System (TAPS). TAPS is being developed under the NASA Turbulence Prediction & Warning System (TPAWS) program. This system generates real-time, automatic reports of hazardous turbulence events and displays the information for improved operations around turbulence. The reports quantify the severity of the loads experienced in the aircraft's cabin in accordance with the standard levels of light, moderate, severe, and extreme as described in the FAA's Airman Information Manual (AIM). These downlinked reports are displayed on dispatchers' flight-following display network, and can be scaled and used to predict and inform other aircraft of potential turbulence encounter severity. Reports are only generated whenever significant turbulence events are encountered. Future efforts will develop the capability to broadcast the reports from aircraft to aircraft and display the reports in the flight deck.

The discussion above describes a reactive solution to turbulence; but if a pilot can get information on the radio that an aircraft ahead is experiencing turbulence, then he can prepare. If there is no warning, then the pilot can only react. Today, technology offers a solution to this problem. Advanced concepts for bird strike warning can be derived from the TAPS system described above. NASA and industry have been trying to solve the turbulence problem for many years, and the TAPS solution is an excellent example of a situation where current technology and existing aircraft data links were used to solve a problem.

To understand TAPS technology, it is important to understand the origin of the turbulence data. Most major airlines participate in the Flight Operational Quality Assurance (FOQA) or Flight Data Management (FDM) Program, whereby digital data recorders are installed on airliners to record flight parameters for the purpose of increasing flight safety. Overseas, many countries are required to participate, but the FAA allows participation on a voluntary basis. The data stored on these recorders is the same as that on the Flight Data Recorders described in Chapter 4. The object of the program is to identify potential problems before an accident takes place. This is especially useful in the case of the flaps, for example. The flaps normally have speed limits associated with each configuration. The greater the flap extension level, the lower the speed limit. When the limit is exceeded, excessive stress can cause a failure in this flight control. Another example is an over-temperature of the engine. If this occurs, engine life will be reduced and a failure could eventually occur. Another application of this device is to record the number of hard landings over a period of time.

Some of the parameters that these flight recorders include are date, time, latitude, longitude, altitude, ground speed, vertical speed, heading, pitch attitude, roll attitude, pitch rate, roll rate, yaw rate, normal acceleration, longitudinal acceleration and lateral acceleration. These and other parameters are required to participate in the FAA program; however, more modern recorders can record up to 500 parameters. The recorders only hold up to 25 hours of flight data, so a FOQA program team consisting of a manager, analyst and pilot typically collect and analyze the data on a routine basis. Application software is used to determine whether any parameter thresholds were exceeded. If this occurs, then corrective actions could be taken, including anything from crew training to changes in operating procedures. Other products can include trend analysis over a one-year period.

The great story behind the TAPS technology is that some bright person was trying to solve the turbulence problem and discovered that there were accelerometers in the recorders described above. TAS uses these accelerometers to record and relay turbulence levels to dispatchers and other aircraft. The TAPS software uses a 5-minute window to determine the RMS (root mean square) value of the z component of acceleration (i.e., normal acceleration up and down). Since 1998, the National Aeronautics and Space Administration (NASA) has been performing research and development on turbulence detection and avoidance systems. The research is in support of their Aviation Safety and Security Program's (AvSSP) who's overall goal is to "develop and demonstrate technologies that contribute to a reduction in aviation accident and fatality rates." NASA and its contractors developed the concepts and initial algorithms of various safety-related technologies under the NASA Turbulence Prediction and Warning System (TPAWS) element of the Weather Accident Prevention (WxAP) Program within AvSSP. The WxAP

Program's three objectives to support the goal of the AvSSP were to develop technologies and methods that will provide pilots with sufficiently accurate, timely, and intuitive information during the en-route phase of flight, which, if implemented, will enable a 25-50% reduction in aircraft accidents attributable to lack of weather situational awareness; to develop communications technologies that will provide a 3- to 5-fold increase in data link system capacity, throughput, and connectivity for disseminating strategic weather information between the flight deck and the ground, which, if implemented along with other supporting technologies, will enable a 25-50% reduction in aircraft accidents attributable to lack of weather situational awareness; and to develop turbulence prediction technologies, hazard metric methods, and mitigation procedures to enable a 25-50% reduction in turbulence-related injuries.

The goals and technical description of the TAPS program below are taken from NASA technical report CR-2007-214887. TAPS was recognized as a technology that was realizable and a significant contributor to meeting the TPAWS goal of "providing airborne centric technologies for detecting and reporting of hazardous turbulence" that when developed would "enable about a 50% reduction in injuries attributable to the lack of turbulence situational awareness." TAPS was further developed and evaluated both in simulations and flight experiments onboard NASA's B757-200 ARIES Research Aircraft. Eventually, the TAPS technology was installed on major airline aircraft for an In-Service Evaluation (ISE). The objective of the TAPS ISE was to develop, implement into commercial aircraft and ground station systems, and evaluate algorithms that would automatically produce reports of aircraft encounters with turbulence and show the reports on ground station displays.

During the TPAWS Program, there was an effort to identify a metric to quantify the turbulence hazard to a commercial transport aircraft. The primary requirements were that it should unambiguously represent the intensity of the turbulence hazard based on accelerations which result in injuries and damage to an aircraft; it should not depend on the atmospheric phenomenon that produce the effect on the aircraft; it could be related to measurements or observables made by various forward-looking airborne sensors (e.g., radar, lidar, etc.); it would be measured by sensors onboard an aircraft, thereby, providing a "truth" measurement to assess the performance of the sensors; and it could be readily scaled from one aircraft to another, based on accepted physics.

The metric that was decided upon was a running 5-second windowed root mean square (RMS) of the vertical acceleration (acceleration parameter). The metric was refined in simulations and several sets of flight experiments on NASA's B757-200 research aircraft under TPAWS. The selection of five seconds was based on two key considerations: the need to balance between a sample window small enough to adequately resolve small-scale turbulence that affect aircraft through induced g-loads, and an accelerometer measurement sample size large enough to calculate an RMS with acceptably low random error.

Analyses were conducted to select thresholds of the acceleration parameter that could be used to define the various levels of turbulence intensity. The selection of thresholds was hampered by the lack of clear, objective data relating the acceleration parameter to the usual subjective descriptions of light, moderate, and severe turbulence. The FAA Airman's Information Manual (AIM) states that during severe turbulence, items in the cabin become weightless. Based on this, a threshold of 0.3g was conservatively chosen as the lower limit of severe turbulence. It is important to note that there is neither a need nor ability to be exact in the categorizations of the turbulence intensities. The need is for thresholds that can be used as a basis for warning pilots and dispatchers of potential hazards, as opposed to attempting a highly accurate scientific quantification of the effect.

TAPS is an automatic system that generates and provides real-time, objective reports of aircraft turbulence encounters in order to improve pilots, dispatchers and air traffic controllers' situational awareness of potential turbulence hazards. TAPS is a combination of non-flight critical software applications residing on an aircraft's computer system and ground station computers that automate the reporting of all significant encounters with turbulence (regardless of convective, clear air, mountain wave classification, or any other source), enable the scaling/interpretation of the turbulence reports for dissimilar aircraft, and display the turbulence information for in the cockpit or for use by ground-based personnel.

The TAPS reporting algorithms consist of three components: hazard metric (acceleration parameter) calculation, reporting logic, and report generation. The hazard metric is continuously calculated during flight. The reporting logic determines when a TAPS report should be generated. When the value of the hazard metric exceeds a particular threshold, a TAPS report is generated containing information on the time, location, aircraft flight conditions, and maximum accelerations experienced during the turbulence encounter. The report also contains a parameter that may be used to scale reports to other aircraft to determine the potential hazard to that particular type of aircraft. An additional benefit of TAPS is that the system would report any encounters when the aircraft exceeded the vertical load acceleration limits, defined by the aircraft maintenance manual, that require an airplane structural examination.

A major component of the architecture of TAPS is to provide the TAPS information to pilots. Multiple TAPS reports can be uplinked automatically to specified aircraft via a ground station in real time. This data link uses ARINC's ACARS messages. ACARS can be transmitted with VHF, HF or SATCOM data links. A turbulence overlay with intensity levels is generated, uplinked and displayed on a cockpit multifunctional display (MFD). Data can also be linked between aircraft using ADS-B.

The main components on the receiving aircraft include a receiving algorithm, which decodes the uplinked TAPS information for use in the interpretation algorithm and return receipt generation; an interpretation algorithm, which scales the received TAPS information and determines if the report represented a potential hazard to the receiving aircraft; and a return receipt packet generation, which combines the TAPS report information and interpretation algorithm results to be downlinked for algorithm verification and validation. The routing rules include the logic by which the ground station can determine whether a TAPS report should be routed to another aircraft within the vicinity. The factors used to make this determination include the range of the aircraft from the location where the TAPS report was generated, the intensity level of the report, and the capability of any particular aircraft to receive the rerouted TAPS report.

ACARS has been around for over 30 years and is the most common aircraft data link used throughout the world. This data link is very slow, though, with transfer rates of only 2400 bps and is normally used for text-based transmission of data. However, image compression algorithms have been developed to convert an image into a text string. So, the conclusion to this story is that existing components (i.e., accelerometers, image compression and cockpit data links) were combined in a novel way with new algorithms to transmit data of critical importance to the cockpit.

Another overlay that is useful is weather. The NEXRAD images can also be compressed and sent to the cockpit using the same architecture. As a matter of fact, several avionics manufacturers have developed and implemented this capability.

With the above as a background, it's time to revisit the current bird strike problem. An advanced concept for sending bird targets to the cockpit would happen as follows. An airport bird radar display could be captured as an image. The same image compression algorithm as was used above could be used to convert the bird snapshot image into a text string. This text string could then be uplinked to the cockpit, decompressed and presented as an overlay on the existing TAPS and NEXRAD displays. In this way, the cockpit could display bird targets in real time., and this real-time bird data could be used for avoidance by the pilot.

A second component to this advanced concept involves NEXRAD. A commercial company has developed algorithms to filter out aircraft, clutter, bats and insects from NEXRAD data. The result is an image with weather and birds. This bird image could be uplinked using the same algorithms and architecture described above. The result would be non-real-time, though, due to infrequent update rates, so its use would have to be thought through. Many times, aircraft are vectored around at low altitudes some 20-30 miles from the airport. One option would be to have the pilot check the bird display at long ranges from the airport to see if birds will be a hazard in the future. It may not localize the problem, but combined with an airport bird radar, it may help identify trends for bird migrations off the airport (i.e., more than 5 miles away) that could become a factor at the airport in the future.

Many times, designed systems find tactical uses other than what was intended in the original design. This is true for TCAS, which was originally designed to be an aircraft collision avoidance system. Once it was deployed and operational, pilots found it to be useful for separation between aircraft as well, which becomes especially critical on final approach to land. If an aircraft gets too close to another, a go-around may be required. The reason for this is that an aircraft cannot land if there is another aircraft on the runway. To prevent the problem from occurring in the first place, the TCAS can be used to ensure proper spacing between aircraft. Normally, you need 2-3 NM between aircraft to avoid a landing conflict. The TCAS display shows the location of all traffic relative to the aircraft. If an aircraft is too close, it can simply slow down to generate a little more spacing. In high-density airspace, this deceleration cannot be accomplished too far out because many aircraft are lined up to land and a deceleration will have a ripple effect on the aircraft behind.

Some engineers are saying "Birds are everywhere, so how could a bird radar be an effective tool for a pilot to avoid hitting a bird?" Well, there are two answers to this question. First, pilots don't see birds everywhere. They program a route of flight into the FMS and fly a fixed heading. Only bird hazards around the projected flight path become a collision hazard. So, that confines the problem. Maybe this means that birds are only a factor when they are in a sector that is +/- 30 degrees of the aircraft heading. This confines the problem and filters it down much more than the "birds everywhere 360 degrees around" scenario, and pilots only need to see what they can hit, not the whole area.

The second answer is that we will not know until we try. It took ten years to develop TCAS and a similar length of time to develop ground collision avoidance systems. It will most likely take ten years to perfect a bird warning and/or collision avoidance system. Today, people are actually afraid to fly because of the airliner in the Hudson. It is only a matter of time before it happens again, so we need to get started on a solution that involves more than a radar that can see birds. We have to develop a means of tactically employing this valuable data so that a pilot can actually make the necessary decisions in time to avoid hitting a bird.

Many engineers are saying the problem is too difficult. What about false alarms? Their flight paths are often unpredictable, so what happens when the bird executes an unexpected flight path? Will the update rates be sufficient for real-time operations? The questions go on and on. However, that is how an engineer is trained to think; all possibilities are considered no matter how low the probability of occurrence is. Worst-case analysis is important, especially in aviation operations. Any time a new component is added to the NAS, it must be done with rigorous analysis and demonstration. Pilots, however, do not share this thinking process. Pilots are trained to take an existing aircraft system and employ it in the most efficient manner. Sometimes this is done for efficiency, and sometimes to prevent a hazardous situation.

An example of employing an aircraft system for efficiency is to use only one engine for taxiing. Most airliners have two engines, so why should you start both at the gate? A fuel savings can be realized if you taxi on one engine and start the other just before takeoff. Sometimes this is not possible, though, if the aircraft is heavy because a very heavy aircraft requires more power. An example of avoiding a hazardous situation is the TCAS avoidance maneuver. Every collision avoidance scenario is different. Pilots are all trained to accomplish the same procedure, but each scenario requires different pilot inputs. One encounter may require a gentle climb and another may require a rapid descent (TCAS maneuvering only allows for climbs and descents, no turns are allowed).

In the military, a similar process takes place. New weapons are always coming out, and pilots need to be trained on their employment. They actually plan missions and use inert weapons (e.g., a weapon with dummy warhead) to train with. You learn procedures and develop habit patterns so that when you go to war, instinct takes over. But before all this takes place, weapons officers have to develop tactics for employment, much in the same way that tactics will need to be developed for the use of bird radars and employing them to avoid birds. Weapons officers are trained at a higher level than other pilots to understand bomb blast and fragmentation patterns as well as what type of bomb works best on a given target. Operational procedures need to be developed to allow for a flight of 2 to 4 aircraft to accomplish a low-level attack at nearly the same time. Once a bomb is dropped, the bomb fragments can fly as high as 2000 feet. You don't want your wingman to fly though the bomb fragments, so you plan a delay or maneuver for the wingman to fly to allow the bomb fragments to hit the ground. In the same fashion, a delay could be used by a pilot to avoid a collision with a bird. If pilots can see the trajectory of a bird flight path, they can plan to avoid it, possibly using a delay to allow the bird targets to fly away from the programmed aircraft flight path.

The bottom line is that we need to get something out there! Pilots will say, "Just give me something I can use." Even if it's an 80% solution, with some problems, that's better than nothing. Right now, pilots are helpless against birds. There are numerous concepts that have been well thought out all the way from design and development to tactical employment by a pilot. They are workable concepts that need to be demonstrated in order to determine feasibility. Stop saying that this is impossible to solve. The same was said about collision avoidance systems and turbulence avoidance. Today, the solutions to those problems are mature. Much research is being done to enhance existing systems and also develop new systems that would be compatible with the NextGen aviation transportation system. We can't just say that NextGen will solve this problem; we need to get started now, perhaps even using the NextGen data links. Use of future technology will ensure survivability. Care must be taken not to use old or current technology that may be phased out like ground-based FAA surveillance radars that will eventually be decommissioned. We need to move away from implementing expensive solutions involving existing FAA surveillance radars.

Chapter 10:

Control Tower Implementation for Bird Radars

Air traffic controllers are responsible for keeping our skies safe, 24 hours a day, 365 days a year. They are responsible for controlling over 60 million aircraft annually and 7,000 aircraft in the air at any given point in time. They also ensure the safety of 600 million passengers per year, which is about 1.6 million passengers per day. This chapter starts by briefly explaining the inner workings of the air traffic control system and air traffic control (ATC) operations (also called 'air traffic management'). Once bird radars are allowed in US control towers, new operational procedures will need to be designed for controllers to help pilots avoid bird strikes, so the final portion of this chapter includes a conceptual description of these procedures.

To become an air traffic controller, one must have education, training and certification. One path to certification is though a four-year Bachelor's of Science Degree program called air traffic management (ATM). Over 20 schools in the United States offer this degree, and many of the instructors have worked as controllers. Normally, these schools have simulators that allow for training, testing and certification, and the graduates of these programs become entry-level ATC managers. A four-year degree helps but is not an absolute requirement. The actual requirement is for an applicant to have either a four-year college degree or three years experience of increasing responsibility. US citizenship is a requirement. The FAA hires school graduates as well as controllers separating from the military. The cutoff age for hiring is 31, and the mandatory retirement age is 56. Currently, in the United States, there are over 14,000 air traffic controllers.

Once an applicant has been selected, the employee attends the FAA Academy in Oklahoma City, OK for 12 weeks of training, where they learn the National Airspace System (NAS). Technical training is provided for both tower and en-route operations. Training starts in the classroom and progresses into simulation using state-of-the-art ATC simulators. After initial training, the controller must pass an exam for an ATC Specialist Certificate. If the controller wants to work in the control tower, he/she must receive further training in basic meteorology, basic air navigation, standard air traffic control and communications procedures, the types and uses of aid to air navigation, and regulations governing air traffic. He must then pass an exam for a Control Tower Operator Certificate. Once the controller is assigned to a control tower, he receives training in the local radio aids to air navigation, the terrain, the landmarks, the communications systems and local area procedures. After this final training, the controller must pass an exam for a rating to operate at the local tower.

The controller's union is called the National Air Traffic Controllers Association (NATCA) and is one of the strongest and most influential labor unions in the federal sector; it is also an affiliate of the American Federation of Labor and Congress of Industrial Organizations (AFL-CIO). NATCA, at over 20,000 members strong, represents not only controllers but other safety-related professionals, including engineers, traffic management coordinators and staff specialists. Air traffic controllers are in high demand. According to NATCA's estimate, nearly half of the nation's controllers will retire by 2010. To combat this huge gap, the union is making every effort possible to ensure adequate staffing in the face of this growing crisis.

Air traffic controllers are federal employees, specifically FAA employees, responsible for the safe separation of aircraft on the ground and in the air. It is a very stressful job with tons of responsibility (i.e., many lives at stake). US Airspace is classified by a lettering system – Class A, B, C, D, E, G and Special Use Airspace. Each airspace classification requires a different level of control and responsibility. To complicate matters even further, every country in the world uses a different structure and classification of airspace and has different rules and regulations. For now, though, let's just stick to US airspace.

Class A airspace is all airspace above the United States from 18000 feet to 60000 feet mean sea level (MSL), which is the height relative to the ocean, as opposed to height above ground level (AGL). The lateral limit of this airspace extends out to 12 NM over the ocean on each coast. Once you start getting closer to an airport, the category of airspace and the type of control changes, depending on traffic density and whether a control tower is present.

Class B airspace looks like an upside down wedding cake, and it is used to keep aircraft under positive ATC control from the surface to 10000 feet. The horizontal limits depend on altitude but generally form concentric circles from the center of the airport to 15 NM at the top tier down to 5 NM at the bottom tier. Examples of airports with Class B airspace include Boston, New York and Atlanta. This airspace is required for the 39 US airports that have greater than 5 million passengers per year. So this is the busiest airspace in the country. To enter this airspace, a pilot must receive clearance from a controller. In this airspace, airliners are normally either climbing or descending, and approach or departure controllers are directing them. In most cases, arrival procedures are assigned by controllers and flown by pilots. There are literally hundreds of these procedures around the country, and they vary from airport to airport. Pilots are required to carry all of these paper procedures with them in their flight kit so that they are prepared to go anywhere at any time. These arrival and departure procedures are designed for efficient traffic flow, noise abatement and, in some cases, to avoid bird migration paths.

If a bird migration path occurs along some portion of the procedure, then the procedure can be changed to avoid birds.

Class C airspace extends around airports that have more than 250,000 passengers per year. This airspace generally extends from the surface to 4000 feet and from the center of the airport out to 5 NM. In the United States, there are 122 Class C airports.

Class D airports are smaller airports with control towers, and controlled airspace generally extends from the surface to 2,500 feet.

Another category of airspace includes Class E, which exists in many forms[21]; Class F, which is not used in the United States; and Special Use, which is mostly reserved for military training operations.

At cruise altitude, airliners normally fly somewhere between 0.8 and 0.9 Mach. Mach 1.0 is the speed of sound, and the actual value changes with temperature. At an altitude of 30000 to 40000 feet, it is somewhere around 600 mph. At the speed of sound, the properties of air change significantly and pressurized shock waves form around the aircraft and in the engines. Airliners actually have structural speed limits that are not to be exceeded or bad things will happen. When an aircraft descends below 10000 feet, there is generally a speed limit of 250 knots (288 mph) in order to reduce the forces of a possible bird strike, and since most bird strikes occur below 3000 feet, the speed limit is further reduced to 200 knots (230 mph) when in Class C or D airspace. The speed limit does not apply in Class B airspace, though, so aircraft can continue to fly at 250 knots.

ATC is normally classified by areas of responsibility. The first is the tower controller who works in the glassed towers that you see at airports. They manage traffic from the airport in a radius of 3 to 30 miles out. They give pilots taxi and takeoff instructions, air traffic clearance, and advice. They provide separation between landing and departing aircraft, transfer control of aircraft to the en-route center controllers when the aircraft leave their airspace, and receive control of aircraft on flights coming into their airspace. Pilots speak with the control tower when they taxi, takeoff and land. The limit of a tower controller's responsibility normally extends to 5 NM. When

21 Some forms include Military Operations Areas (MOAs), Warning Areas and Restricted airspace, which are used primarily for military operations.

an aircraft flies outside this range, control is normally transferred to a terminal controller.

The second area of responsibility is the terminal radar approach controller. These controllers work in rooms with radar displays, usually in airport towers. They are responsible for the safe separation and movement of aircraft departing, landing, and maneuvering in the airport environment. They normally control traffic below 17000 feet. On departure, the aircraft is climbing to a final altitude, which is normally above 30000 feet on longer flights. This altitude is chosen because it is most efficient for jet engines and conserves fuel. As the aircraft climbs to a higher altitude, the final controller directs en-route traffic. On arrival, as the aircraft is descending, the controller's job is to direct aircraft to the airport as efficiently as possible.

The final area of responsibility is the en-route center controller. These controllers work in 24 centers across the country at a location far away from the airport. Each center has a controller. As an aircraft transitions from one center to the next, he has to change frequency and talk to another controller in the next center's airspace.

While flying in high-altitude airspace and transitioning between sectors, an aircraft will be handed off to the next controller. As you can see, there are several areas an aircraft transitions through, and a transition between areas requires a handoff to a new controller. Most of the flight occurs in center airspace. The area of responsibility of a center controller is aircraft above 17000 feet. These controllers give aircraft instructions, air traffic clearances, and advice regarding flight conditions during the en-route portions of flights. Center controllers use radar or manual procedures to keep track of the thousands of planes in the sky at any one time. The typical center has responsibility for more than 100,000 square miles of airspace generally extending over a number of states.

A standard terminology is used when pilots and controllers communicate, and an air traffic controller's glossary of terminology must be followed exactly to prevent confusion as to the meaning of a controller's instructions. Many accidents have been caused by the use of non-standard terminology by either the pilot or controller. Currently, this communication takes place over a radio, but several aircraft are normally tuned to the same frequency. As the number of aircraft increases, the number of conversations that occur increases and, at some point, the radios become saturated, which makes communications difficult; however, new technologies that use data links are being tested that allow pilots and controllers to communicate using text messages, which reduces congestion on the radio.

Worldwide, the standard for communication is English, and controllers in other countries need English language proficiency; however, I have experienced difficulties understanding the accents of air traffic controllers in many countries. Because of the accent, I've had to ask a controller to repeat instructions for clarification. If I still did not understand after the repeat, I asked the controller to give the phonetic spelling. For example, say the controller cleared me direct to Frankfurt (FFA) and I did not understand him. If I ask, then the controller would spell this as "Fox Fox Alpha." To avoid a safety issue, it is extremely important to get clarification from controllers in the case of a potential misunderstanding.

The worst aviation disaster in history, in fact, occurred due to improper communications. On 27 March, 1977, in one of the Canary Islands, 583 people died when an aircraft taking off collided with another aircraft on the ground. It was a foggy day, and the control tower cleared one aircraft to back taxi (i.e., taxiing in the direction opposite to that of takeoff). Another aircraft at the end of the runway was supposed to stay put until the back taxiing aircraft was clear of the runway. The aircraft at the end of the runway misunderstood the controller's instructions and thought he heard clearance to take off. So the pilot took off and clipped the top of the other aircraft, causing a crash and a significant loss of life. This accident had several causes, including non-standard terminology between the pilot and controller. So, correct, precise, standardized terminology is essential for safe operations.

The controller's job is to keep aircraft separated in order to prevent collisions and make the flow of traffic as efficient as possible in order to conserve fuel. In addition to directing traffic, controllers also provide information on winds, weather and turbulence. While you are riding as a passenger, the pilot is actually talking to the controller, asking where there is a smooth altitude. He does this by asking aircraft in front if they are at a smooth altitude. Once a smooth altitude is discovered, at the pilot's request, the controller clears the aircraft to a different altitude above or below its present altitude. Pilots and controllers are always trying to provide a smooth ride; however, there are some conditions where smooth rides are not possible, and all must endure temporarily.

Basically, controllers face two major problems: overcapacity and weather. When an aircraft lands, it must slow down and clear the runway, which takes a certain amount of time. At some airports, the controller may also need to sequence both takeoffs and landings on the same runway. At busy airports, as I mentioned earlier, there are operations counts of more than one every 15 seconds. To accommodate such high operations counts, several parallel runways with simultaneous operations are sometimes necessary. Problems occur, though, when the airlines schedule more arrivals into an airport than a controller can handle. This causes delays. While traveling as a passenger, you may have heard the pilot come on the intercom and say that you will be doing a turn in holding due to ATC delays. Holding is often needed in order to sequence the overcapacity aircraft into the controller's airspace. In recent history, air traffic management has made significant advances. Computer application software allows for the sequencing of aircraft many hours in advance of the takeoff. As a result, aircraft are sometimes given a slot time (i.e., a time interval for takeoff) to accommodate the delays so that the aircraft does not need to hold and waste fuel at the destination.

The second problem a controller faces is the weather. Ice, rain and snow are major obstacles to operations. When these types of precipitation occur on runways, it will take longer for aircraft to slow down and exit the runway. This can cause excessive delays because the rate for aircraft arrivals will decrease and require more spacing between aircraft. This has the ripple effect of causing more aircraft to enter a holding pattern. If the delays for weather become excessive, then a ground hold may be implemented to avoid wasting gas in holding at the destination. Weather also creates difficulties for controllers in the form of thunderstorms. Since thunderstorms can tower up to altitudes in excess of 40000 feet, they become major obstacles in getting to the airport. Pilots need to deviate above, around or through these storms which will cause delays and reduce capacity.

To handle these two problems on a national level, an office called the ATC System Command Center (ATCSCC) was established in Washington, DC. When weather, equipment, runway closures or other situations cause an impact to the NAS, this office regulates air traffic. In order to prevent overcapacity, the ATCSCC works with controller and airline personnel to make modifications to traffic flow across the United States. At the ATCSCC, there are several offices. The first is the Airport Reservation Office (ARO), which processes all flights at JFK, LaGuardia and Reagan National airports. This office handles traffic when there is high demand at one of these airports due to special events like the Olympics or a golf tournament. By implementing the Special Traffic Management Program (STMP), they control the number of arrivals and departures in order to limit the number of reservations over a specific period of time.

The Enhanced Traffic Management System (ETMS) is used to predict, on national and local scales, traffic surges, gaps, and volume based on current and anticipated airborne aircraft. Controllers evaluate the projected flow of traffic into airports and sectors, then implement the least restrictive action necessary to ensure that traffic demand does not exceed system capacity. A part of ETMS analyzes traffic demand for all airports, sectors, and airborne reporting fixes[22] in the continental United States, and then automatically displays an alert when demand is predicted to exceed capacity in a particular area. Armed with this information, controllers examine the situation, and then provide routes and spacing to assist in controlling the flow of traffic around the United States.

Another ATCSCC office is the National Operations Control Center (NOCC). The NOCC was established to provide national coordination of facility restoration activities and to provide status information on NAS equipment. The NOCC is staffed 24 hours a day, 7 days a week, and is responsible for collecting, tracking and reporting data from field organizations for NAS status, facility

22 An airborne reporting fix is required when a pilot's loss of radar contact requires him to report his position via radio.

and service interruptions, special events and disasters. The NOCC monitors critical situations as they evolve and notifies, mobilizes or directs key organizations while coordinating these events with air traffic and airway facilities. The NOCC is a conduit to the regional Maintenance Control Centers (MCC) for restoration of key NAS facilities and services, and communicates electronically with FAA technicians responsible for maintaining the NAS.

Another ATCSCC office is the National System Strategy Team (NSST), which manages the NAS when bad weather develops in order to minimize the impact of severe weather on the airspace system. The NSST collects meteorological information from a variety of sources and devises a suitable plan with other air traffic facilities and system users for routing traffic around the bad weather. Through coordination with affected air traffic facilities and users, an operational plan is developed to ensure the safe and orderly flow of air traffic around impacted areas. As the weather changes, the plan can be revised, and, if necessary, new routings or restrictions are issued.

Now, let's take a closer look at where bird radars could be installed – control towers at airports. The equipment in a control tower includes radios for communications with aircraft. A controller usually has a headset and microphone to talk to the pilots. There are also telephones, which are used to talk to other controllers in airports or centers. If the communications at an airport goes down, portable VHF radios are sometimes used. Failing this, there is a light gun that is aimed at the aircraft to communicate various clearances and conditions. There are also weather gauges that present information, such as pressure and temperature and winds. A radar display is used to control aircraft within 5 NM of the airport. It is also important to note that controllers use their eyes to control aircraft. Radar is helpful when aircraft are outside visual range, but when a controller acquires an aircraft visually, he normally uses the naked eye for safe separation between aircraft. Full control tower structures usually have windows that circle the entire top floor, giving all-around vision. It is interesting to note that the windows in a control tower are tilted at an angle. The reason for this tilt is to reduce the glare on control tower displays.

Most airports in the world do not have control towers, and no tower means no air traffic controller. Since there is no controller, there must be another means of sequencing traffic. This occurs through standard procedures, which include radio calls to report aircraft position. Airports without control towers may exist below Class B airspace. If this is the case, even if there is no control tower, pilots are required to obtain clearance to takeoff from the airspace controller. These clearances do not have to occur on the radio; phone calls to various agencies can be used to obtain clearance. When an aircraft enters the traffic pattern, a common local frequency is normally established to communicate aircraft position within the traffic pattern. A radio call with aircraft position allows an aircraft to visualize where other aircraft are relative to their current position.

Airports without towers can be dangerous places. If pilots forget to report their positions, a potential for collision exists because pilots will not know where other aircraft are in the traffic pattern. As time goes by, if the airport becomes popular and the number of operations increases, the airport may be considered for a control tower. Then, the FAA has to find the money to build the tower and pay for the controller. For some events, a control tower may be established on a temporary basis. One example of this is Whittman Field in Oshkosh, WI. This airport hosts an annual fly-in and becomes the world's largest airport for a week. Many operations, including ATC, are conducted on a volunteer basis.

Much has changed over the years. Early commercial flights were restricted to daytime-only operations. During the 1920s, aircraft used to follow a path on the ground that used gas fires or bonfires. After that, pilots would fly over ground-based landmarks and report their position to a ground-based controller. They would report their location, altitude, time and ETA to the next point. Speed control would ensure aircraft spacing when aircraft were at the same altitude. The controller used a blackboard and a map to keep track of aircraft. To mark an aircraft's position, they used markers on a map, and these markers were called "shrimp boats" because of their shape. In fact, these position reports are still used today for places where there is no radar coverage. I've had this situation happen to me – while flying over the ocean along the southern coast of the United States; I have lost radar coverage and had to report my position via the radio. These position reports are also required when flying over the ocean when an aircraft is not in radar contact.

Radar is the primary means of detecting and tracking aircraft. FAA radars are installed at airports and around the country at various locations so that aircraft can be controlled anywhere in the United States. The radar can process the aircraft detection and compute an aircraft heading and speed. A secondary role of these radars is to provide aircraft locations to the military (NORAD) in order to defend US airspace against foreign intruders. Another device used by ATC to observe an aircraft's position altitude and code is called a transponder (sometimes called 'secondary surveillance radar'). A radar can only see an aircraft – it cannot identify it. An aircraft transponder, however, adds ID and altitude capability so that the controller knows who's who. There is a transponder interrogator at the airport that sends an interrogation signal. The aircraft receives this signal and sends a reply to the controller. The radar target and the interrogation of an aircraft's transponder are fused together into a composite track with the aircraft code, altitude and heading and are placed on a display at a position relative to the radar. Bird target returns are normally processed and filtered out. They have always been thought of as noise; no one ever thought we would actually want to see them in the data or on the display.

A bird radar has limited range and will be employed at the airport only, so it is important to discuss bird radar operations in the control tower. The first area of responsibility for control tower operations is ground control. When an aircraft is taxiing to the runway, it is controlled by a ground controller. In addition to aircraft, the control tower is responsible for anything that moves on the airport surface. Control happens though radio transmission. Some very busy airports have ground-based radars that can see anything that moves and place a vehicle or aircraft radar target on a display. So there are separate radars for ground and airborne operations.

The tower is responsible for clearing aircraft to takeoff and land and is also responsible for safe separation of aircraft. If you have traveled a lot by plane in the past, at some point in time, your pilot was probably getting ready to land when, you suddenly felt an acceleration and the aircraft pitched up and went back around the traffic pattern. A good pilot will come on the intercom and tell you why. Most likely, what happened is that an aircraft was on the runway when you were trying to land. The runway must be clear before an aircraft can land (though this is not true for military operations). Just to clarify, the controllers that control airborne traffic are separate from ground controllers. These controllers, however, have to be in very close communication. For example, an aircraft is attempting to land and another on the ground wants to cross the runway. The ground controller must get permission from the airborne controller before he clears the aircraft to cross. If he doesn't do his job correctly, or if there is an excessive delay, then there could be collisions between an airborne aircraft and an aircraft on the ground.

Another position in the control tower is clearance delivery. Before a pilot starts to taxi, he must call the controller and get clearance for the aircraft route of flight all the way to the destination. A flight plan is normally filed, and clearance delivery normally tries to accommodate the plan, but changes occasionally happen in order to accommodate other traffic. The tower must normally coordinate with the en-route centers described above in order to obtain a release. Sometimes, if the airspace is full or the weather is bad, a delay will occur before the release is approved. I am sure if you travel a lot, you have heard the pilot announce the reason for the delay being either weather or air traffic control. ATC means that the airspace is full and must clear out before another aircraft can be accommodated.

Bird radars (normally called 'ornithological' or 'avian radars') are typically used to conduct bird and bat surveys or to monitor wind farms, landfills, oil and gas facilities, power plants, mines, and communication towers. Recently, though, these avian radars have been adapted for airport use. Bird targets are displayed on an airport overlay for situational awareness. Various signal processing methods are used to determine how many birds are present. Success has been limited, though, in integrating these radars into control tower operations. The primary reason for this is that in the US civil sector, safety is a primary consideration. Any equipment added to an airport environment must go through years of rigorous testing to ensure that safety is not compromised. In addition to technical analysis and testing, the new systems must be operationally useful to controllers. Operational suitability is normally determined by testing at an airport with controller involvement. The testing is accomplished at several airports, and controllers are asked to fill out surveys. Once this process is complete, the new piece of equipment is added to an FAA-approved list.

In my opinion, there are three primary users for data from a bird radar: wildlife technicians, controllers and pilots. The operations are different for each set of users, but since the controller is the primary focus of this chapter, let us focus now on ATC use of avian radar. Before a controller can use a bird radar, the airport has to purchase one. When an airport makes the decision to purchase a new piece of equipment, they normally go out for quotes from FAA-approved vendors. Once the quotes are received, the airport management approves the request and a purchase order is issued. Currently, the FAA is evaluating several bird radars, and I assume that, at the end of the evaluation, some vendors will be approved for airport use. Once this occurs, airports will be able to have permanent bird radars installed, and they will be approved for use in the control tower; however, bird radars are not currently approved for use in US civil control towers.

In Seattle, Chicago and New York, experimental bird radars have been installed by the FAA and are currently going through an evaluation process, but these are the wrong places to start an evaluation of bird radars for operational use in a control tower. One reason is that these airports are all Class B with severe capacity constraints. When operations counts exceed 240 per hour, as they do in these airports, there is no time for controllers to look at a bird radar display. A statistic of 240 operations per hour translates to one operation every 15 seconds (an 'operation' is either a takeoff or landing). Furthermore, because so many aircraft are trying to use the same frequency at these airports, there are also severe limitations on the use of the radio to communicate. I have been on flights where I've had to wait a long time to make a radio transmission because there were so many pilots and controllers talking at the same time. One reason for this radio saturation is that when controllers issue instructions to pilots, pilots have to repeat the instructions back.

This way, if there is an error in the pilot's understanding, then the controller can correct the pilot and reissue the instruction.

In-flight misunderstandings are even more critical than on the ground. The third deadliest air crash and the worst mid-air collision in aviation history occurred in November 1996. A Saudi Arabian jumbo jet collided after takeoff with a Kazak Airlines plane approaching the New Delhi airport, and 315 passengers and crew were killed. The en-route controller cleared one aircraft to descend to 15000 feet and cleared the other aircraft, which was going in the opposite direction, to climb to 14000 feet. The pilot of the aircraft that was directed to descend to 15000 feet misunderstood the controller and descended to 14000 feet. This misunderstanding caused a head-on collision of the two aircraft. So, you see why it is important for pilots to read back the controller's instructions.

Control towers at the busy Class B airports have no time to look at a separate bird radar display, but even if they did, frequency congestion would make it difficult to communicate this information to the pilot. Add to this the fact that there are only 39 Class B airports and you can see the limited utility of conducting bird radar evaluations at these airports. Another problem with using bird radars in control towers at Class B airports is the delay in command and control. By the time the controller sees a bird conflict and relays it to the pilot, it may be too late for any action to be taken to resolve the conflict.

The use of a bird radar in control towers is a recent development over the last three years. According to my research, bird radars are currently used in the US military, the UK Royal Air Force, an African civil control tower, and, as of the writing of this book, a bird radar was sold to Bagram Airbase in Afghanistan, the first combat base to use a bird radar in a control tower. You may ask why the bird radars are successful in these places and why they can't be just as successful at busy Class B airports. Again, the answer lies in the operations count. I checked the operations counts at most military air bases, and they are on the order of one takeoff and landing per hour. With such a low operations count, the controller would have all day to look at a bird radar display and discuss hazardous bird activity on the radio with a pilot. In my opinion, a bird radar would have the most success in military and civil control towers at airports with low operations counts. My recommendation is that any government evaluation of a bird radar should be conducted in a lower-traffic airport in order to give the tower controller time to assess the display and talk to the pilot. The current evaluation at Class B airports could fail due to the controller's lack of extra time.

Tests have been conducted at US control towers, and the feedback given was that the bird radar requires automation so that the controller does not have to stare at a display while performing so many other tasks. In such a fast-paced environment, controllers need to look at a display and get information in the shortest amount of time possible, so if this bird radar is to be installed at high-density Class B airports, then automation is clearly essential. One method of automation is to have an audible warning whenever the bird activity level changes. A radar can determine how many birds are present and assess a risk level. More birds mean a higher level of risk. Thresholds can be established for low, medium and high levels of bird activity. When the threshold changes, then an audible warning could sound, signaling the controller to look at a display and determine where the activity is occurring. The airport could be divided up into sectors around the runway. In conjunction with the audible warning, color coding could be established to indicate the various risk levels by sector. Low could be green; medium could be yellow and high could be red.

Currently, bird radar displays are separate from the radar displays used to control traffic. In most cases, there is one display for horizontal scan coverage and another for vertical scan coverage over a narrow area aligned with the runway. It is tedious, though, to have to look at three separate displays in order to determine whether birds are a hazard, so at some time in the future, the bird targets should be integrated with current air traffic control displays so that a common picture can be presented. Most controllers would probably object to a more cluttered display, but this can be managed in a number of ways. Suppose that the bird targets are integrated into the controller's display, and the controller uses a switch to turn the bird targets off and on. Whenever there is a warning, the controller could turn the switch on to display the bird targets. Then, once he sees where the birds are located, the switch could be turned off to declutter the screen.

The next several chapters discuss how a team of aviation professional's work together to accomplish a successful flight and why bird warnings need to be automated in the cockpit. After that, the optimal solution of sending the bird targets directly to the cockpit is discussed, eliminating the need for a controller to look at an additional display, interpret the data and make a transmission to a pilot. All of these actions will cause a delay, and by the time a controller gets on the radio to tell the pilot where the conflict is, the birds will probably no longer be at the same position. One must also consider that there will be a certain time delay each time the pilot has to interpret the cockpit display and take action. At very high jet speeds, delays simply cannot be tolerated. This is why the optimal solution would be to send the bird targets directly to the cockpit. This cuts the controller and the associated delays completely out of the picture and enables pilots to perform an immediate avoidance maneuver that directly enhances safety.

In summary, this chapter presents a brief description of how air traffic controllers have kept our skies safe. Several implementations that depend on the number of operations at an airport were presented; high operations counts require more automation and low operations counts can tolerate the time it takes for a controller to observe and interpret a bird radar display. An optimal solution would be to send the bird targets simultaneously to both the pilot and the controller, using the same data link and the same display. This way, both the pilot and controller can see the same picture at the same time. Both the pilot and controller could thereby be on the lookout for hazardous conditions. Automated warnings, as explained in Chapter 15, could be installed in both control towers and cockpits to warn both controllers and pilots of potential bird collisions, thereby giving the pilot slightly more time to react.

Chapter 11:

Teamwork in Aviation

In order for a flight to be conducted safely, a team of aviation industry experts must work together flawlessly. This collaborative group effort involves the efforts of many subgroups and individuals. In the cockpit, two pilots work together as a team. Airspace is divided up into sectors, and there is a controller in each sector. Controllers work together to guide an aircraft through all the different types of airspace. Dispatcher teams work in the background filing flight plans; checking the weather; determining how much fuel must be loaded onto the aircraft; making weight and balance computations for passengers, cargo and fuel; ensuring that proper maintenance is conducted; coordinating actions when emergencies occur and maintaining contact with the crew while the flight is in the air. Finally, the maintenance team performs required maintenance to ensure that the aircraft is safe to fly. An error made by any one of these individuals can cause severe disasters and significant loss of life. This chapter will demonstrate the magnitude of the consequences that have occurred when one of these teams has failed to do their job.

As you read through these examples of aviation teamwork gone wrong, please keep in mind that successful resolution to the aircraft bird strike problem will also require a team. Teamwork will be needed to design, develop, integrate, test, maintain, operate and improve technical solutions to the bird strike problem. This team will include pilots, controllers, wildlife biologists, wildlife control specialists, the FAA, the USDA, the EPA, the DoD, airport management and airport operations. During daily operations, the tactical aviation team that will be involved in real-time bird strike avoidance will consist of pilots, controllers and wildlife specialists. By examining situations where teamwork has fallen apart, you will develop a stronger appreciation for the important role that each individual plays in making the optimal solution possible.

In 1985, the worst single aircraft aviation accident in history occurred due to a breakdown in the maintenance team. Shortly after takeoff, a Japanese 747 suffered a mechanical failure and crashed into the mountains, causing the death of 520 passengers and crew. After the aircraft had reached its cruise altitude, a decompression (and the resulting explosion) tore off the vertical stabilizer and severed all four of the aircraft hydraulic systems. Hydraulics are required to operate the flight controls. After the hydraulic fluid had drained out, the aircraft became difficult to control. Initially, the pilots were able to use engine power to cause the aircraft to climb and descend. They used differential thrust to turn the aircraft. They were able to maintain control for a short period of time, but control was eventually lost, and the aircraft plunged into a mountain. The investigation revealed that several years earlier, the aircraft had been involved in a tail strike. A tail strike can occur on either takeoff or landing when the nose of the aircraft is too high and the tail touches the runway. After the tail strike, the maintenance team had riveted a metal plate over the damaged area; however, they had failed to follow Boeing procedures for properly installing the metal plate. Over several years, the plate had become fatigued and ultimately failed.

Another example of a maintenance team failure occurred in 1974 on a Turkish DC-10. Until the aforementioned accident happened in Japan, this was the worst single aircraft aviation accident in history with 346 deaths. A rear cargo hatch blew off during flight, which caused a rapid decompression. The decompression caused the cabin floor to collapse, which severed the flight control cables and caused the aircraft to go out of control. The investigation revealed that the maintenance team had not made mandatory changes to the latching mechanism. The changes were never implemented, and it is not known whether this was a deliberate act or oversight.

In 1991, a controller team failure led to 33 deaths when a 737 landed on top of a commuter aircraft in position for takeoff at Los Angeles International Airport. The controller told the commuter aircraft to wait on the runway and, several seconds later, cleared the 737 to land.

In 2001, a controller team failure led to the crash of two F-15s in England. During a snowstorm, a controller directed the F-15s to descend to an altitude that was too low for the terrain. The result was a collision with a mountain.

In 1978, a pilot crew team failure caused the crash of an airliner in Portland, OR after one of the landing gear would not come down. While the crew was focusing on solving the problem, they forgot to calculate how much fuel they had left. Ultimately, they ran out of fuel and crashed, killing 10 passengers. This crash occurred because of a breakdown in communication between members of the cockpit team.

After numerous accidents of the types described above, the FAA started to investigate the causes and discovered that in 60 to 80% of air carrier accidents, human error was listed as a contributing factor. Problems were attributed to poor group decision making, ineffective communication, inadequate leadership, and poor task or resource management. Pilot training programs historically focused almost exclusively on the technical aspects of flying and on an individual pilot's performance; they did not effectively address the crew management issues that are also fundamental to safe flight. Industry and the government have come to the consensus that training programs should place emphasis on the factors that influence crew coordination and the management of crew resources. The need for additional training in communication between cockpit crewmembers and flight attendants was specifically identified. The result of this study was Advisory Circular 120-61E on Cockpit Resource Management Training, which recommended that pilots, controllers, dispatchers and flight attendants have a CRM program incorporated into their training. The following is an excerpt from the AC:

CRM training is based on awareness that a high degree of technical proficiency is essential for safe and efficient operations. Demonstrated mastery of CRM concepts cannot overcome a lack of proficiency. Similarly, high technical proficiency cannot guarantee safe operations in the absence of effective crew coordination. Experience has shown that lasting behavior changes in any environment cannot be achieved in a short time, even if the training is well designed. Trainees need awareness, practice and feedback, and continuing reinforcement: in brief, time to learn attitudes and behaviors that will endure. To be effective, CRM concepts must be permanently integrated into all aspects of training and operations. While there are various useful methods in use in CRM training today, certain essentials are universal: CRM training is most effective within a training program centered on clear, comprehensive SOPs; CRM training should focus on the functioning of crewmembers as teams, not as a collection of technically competent individuals; CRM training should instruct crewmembers how to behave in ways that foster crew effectiveness; CRM training should provide opportunities for crewmembers to practice the skills necessary to be effective team leaders and team members; CRM training exercises should include all crewmembers functioning in the same roles (e.g., captain, first officer, and/or flight engineer, flight attendants) that they normally perform in flight; CRM training should include effective team behaviors during normal, routine operations.

Good training for routine operations can have a strong positive effect on how well individuals function during times of high workload or high stress. During emergency situations, it is highly unlikely (and probably undesirable) that any crewmember would take the time to reflect upon his or her CRM training in order to choose the appropriate behavior. But practice of desirable behaviors during times of low stress increases the likelihood that emergencies will be handled effectively. Effective CRM has the following characteristics: CRM is a comprehensive system of applying human factors concepts to improve crew performance; CRM embraces all operational personnel; CRM can be blended into all forms of aircrew training; CRM concentrates on crewmembers' attitudes and behaviors and their impact on safety; CRM uses the crew as the unit of training; CRM is training that requires the active participation of all crewmembers. It provides an opportunity for individuals and crews to examine their own behavior, and to make decisions on how to improve cockpit teamwork.

Research programs and airline operational experience suggest that the greatest benefits are achieved by adhering to the following practices:

Assess the Status of the Organization Before Implementation. It is important to know how widely CRM concepts are understood and practiced before designing specific training. Surveys of crewmembers, management, training, and standards personnel, observation of crews in line observations, and analysis of incident/accident reports can provide essential data for program designers.

Get Commitment from All Managers, Starting with Senior Managers. CRM programs are received much more positively by operations personnel when senior managers, flight operations managers, and flight standards officers conspicuously support CRM concepts and provide the necessary resources for training. Flight operations manuals and training manuals should embrace CRM concepts by providing crews with necessary policy and procedures guidance centered on clear, comprehensive SOPs. A central CRM concept is communication. It is essential that every level of management support a safety culture in which communication is promoted by encouraging appropriate questioning. It should be made perfectly clear in pilots' manuals, and in every phase of pilot training, that appropriate questioning is encouraged and that there will be no negative repercussions for appropriate questioning of one pilot's decision or action by another pilot.

Customize the Training to Reflect the Nature and Needs of the Organization. Using knowledge of the state of the organization, priorities should be established for topics to be covered, including special issues, such as the effects of mergers or the introduction of advanced technology aircraft. Other special issues might include topics specific to the particular type of operation, such as the specific characteristics that exist in commuter operations, in long-haul international operations or night operations. This approach increases the relevance of training for crewmembers.

Define the Scope of the Program and an Implementation Plan. Institute special CRM training for key personnel, including check airmen, supervisors, and instructors. It is highly beneficial to provide training for these groups before beginning training for crewmembers. CRM training may be expanded to combine pilots, flight attendants, and aircraft dispatchers. It may also be expanded to include maintenance personnel and other company team members, as appropriate. It is also helpful to develop a long-term strategy for program implementation.

Communicate the Nature and Scope of the Program Before Startup. Training departments should provide crews, managers, training, and standards personnel with a preview of what the training will involve together with plans for initial and continuing training. These steps can prevent misunderstandings about the focus of the training or any aspect of its implementation.

Institute Quality Control Procedures. It has proved helpful to monitor the delivery of training and to determine areas where training can be strengthened. Monitoring can be initiated by providing special training to program instructors (often called 'facilitators') in using surveys to collect systematic feedback from participants in the training.

The topics outlined below have been identified as critical components of effective CRM training. They do not represent a fixed sequence of phases, each with a beginning and an end. Ideally, each component is continually renewed at every stage of training.

Initial Indoctrination/Awareness. Indoctrination/awareness typically consists of classroom presentations and focuses on communications and decision making, interpersonal relations, crew coordination, leadership, and adherence to SOPs, among others. In this component of CRM training, the concepts are developed, defined, and related to the safety of line operations. This component also provides a common conceptual framework and a common vocabulary for identifying crew coordination problems. Indoctrination/awareness can be accomplished by a combination of training methods. Lectures, audiovisual presentations, discussion groups, role-playing exercises, computer-based instruction, and videotaped examples of good and poor team behavior are commonly used methods. Initiating indoctrination/awareness training requires the development of a curriculum that addresses CRM skills that have been demonstrated to influence crew performance. To be most effective, the curriculum should define the concepts involved and relate them directly to operational issues that crews encounter. Many organizations have found it useful to survey crewmembers. Survey data have helped identify embedded attitudes regarding crew coordination and cockpit management. The data have also helped to identify operational problems and to prioritize training issues. Effective indoctrination/awareness training increases understanding of CRM concepts. That understanding, in turn, often influences individual attitudes favorably regarding human factors issues. Often, the training also suggests more effective communication practices. It is important to recognize that classroom instruction alone does not fundamentally alter crewmember attitudes over the long term. The indoctrination/awareness training should be regarded as a necessary first step towards effective crew performance training.

Recurrent Practice and Feedback. CRM training must be included as a regular part of the recurrent training requirement. Recurrent CRM training should include classroom or briefing room refresher training to review and amplify CRM components, followed by practice and feedback exercises, preferably with taped feedback; or a suitable substitute, such as role-playing in a flight training device and taped feedback. It is recommended that these recurrent CRM exercises take place with a full crew, each member operating in his or her normal crew position. A complete crew should always be scheduled, and every attempt should be made to maintain crew integrity. Recurrent training, which includes CRM, should be conducted with current line crews, and preferably not with instructors or check airmen as stand-ins. Recurrent training with performance feedback allows participants to practice newly improved CRM skills and to receive feedback on their effectiveness. Feedback has its greatest impact when it comes from self-critique and from peers, together with guidance from a facilitator with special training in assessment and debriefing techniques. The most effective feedback refers to the coordination concepts identified in Indoctrination/Awareness training or in recurrent training. Effective feedback relates to specific behaviors. Practice and feedback are best accomplished through the use of simulators or training devices and videotape. Taped feedback, with the guidance of a facilitator, is particularly effective because it allows participants to view themselves from a third-person perspective. This view is especially compelling in that strengths and weaknesses are captured on tape and vividly displayed. Stop action, replay, and slow-motion are some of the playback features available during debriefing. Behavioral patterns and individual work styles are easily seen, and appropriate adjustments are often self-evident.

Continuing Reinforcement. No matter how effective each curriculum segment is (the classroom, the role-playing exercises . . . or the feedback, one-time exposures are simply not sufficient. The attitudes and norms that contribute to ineffective crew coordination may have developed over a crewmember's lifetime. It is unrealistic to expect a short training program to reverse years of habits. To be maximally effective, CRM should be embedded in every stage of training, and CRM concepts should be stressed in line operations as well. CRM should become an inseparable part of the organization's culture. There is a common tendency to think of CRM as training only for captains. This notion misses the essence of the CRM training mission: the prevention of crew-related accidents. CRM training works best in the context of the entire crew. Training exercises are most effective if all crewmembers work together and learn together. In the past, much of the flight crew training has been segmented by crew position. This segmentation has been effective for meeting certain training needs, such as seat dependent technical training and upgrade training, but segmentation is not appropriate for most CRM training. Reinforcement can be accomplished in many areas. Training such as joint cabin and cockpit crew training in security can deal with many human factors issues. Joint training with aircraft dispatchers, maintenance personnel, and gate agents can also reinforce CRM concepts, and is recommended.

The topics outlined below have been included in many current CRM programs. Specific content of training and organization of topics should reflect an organization's unique culture and specific needs.

Communications Processes and Decision Behavior. This topic includes internal and external influences on interpersonal communications. External factors include communication barriers, such as rank, age, gender, and organizational culture, including the identification of inadequate SOPs. Internal factors include speaking skills, listening skills and decision-making skills, conflict resolution techniques, and the use of appropriate assertiveness and advocacy. The importance of clear and unambiguous communication must be stressed in all training activities involving pilots, flight attendants, and aircraft dispatchers. The greater one's concern in flight-related matters, the greater is the need for clear communication.

Team Building and Maintenance. This topic includes interpersonal relationships and practices. Effective leadership/followership and interpersonal relationships are key concepts to be stressed. Curricula can also include recognizing and dealing with diverse personalities and operating styles. Subtopics include:

Leadership/Followership/Concern for Task. Showing the benefits of the practice of effective leadership through coordinating activities and maintaining proper balance between respecting authority and practicing assertiveness. Staying centered on the goals of safe and efficient operations.

Interpersonal Relationships / Group Climate. Demonstrating the usefulness of showing sensitivity to other crewmembers' personalities and styles. Emphasizing the value of maintaining a friendly, relaxed, and supportive yet task-oriented tone in the cockpit and aircraft cabin. The importance of recognizing symptoms of fatigue and stress, and taking appropriate action.

Workload Management and Situation Awareness. Stressing the importance of maintaining awareness of the operational environment and anticipating contingencies. Instruction may address practices (e.g., vigilance, planning and time management, prioritizing tasks, and avoiding distractions) that result in higher levels of situation awareness.

Individual Factors / Stress Reduction. Training in this area may include describing and demonstrating individual characteristics that can influence crew effectiveness. Research has shown that many crewmembers are unfamiliar with the negative effects of stress and fatigue on individual cognitive functions and team performance. Training may include a review of scientific evidence on fatigue and stress and their effects on performance. The content may include specific effects of fatigue and stress in potential emergency situations. The effects of personal and interpersonal problems and the increased importance of effective interpersonal communications under stressful conditions may also be addressed. Training may also include familiarization with various countermeasures for coping with stressors. Additional curriculum topics may include examination of personality and motivation characteristics, self-assessment of personal style, and identifying cognitive factors that influence perception and decision making.

Other Types of CRM Training

Crew Monitoring and Cross-Checking. Several studies of crew performance, incidents, and accidents have identified inadequate flight crew monitoring and cross-checking as a problem for aviation safety. Therefore, to ensure the highest levels of safety, each flight crewmember must carefully monitor the aircraft's flight path and systems and actively cross-check the actions of other crewmembers. Effective monitoring and cross-checking can be the last line of defense that prevents an accident because detecting an error or unsafe situation may break the chain of events leading to an accident. This monitoring function is always essential, and particularly so during approach and landing when controlled flight into terrain (CFIT) accidents are most common.

Joint CRM Training. More carriers are discovering the value of expanding CRM training to reach various employee groups beyond flight crew and flight attendants. Dissimilar groups are being brought together in CRM training and in other activities. The objective is to improve the effectiveness and safety of the entire operations team as a working system. The attacks of September 11, 2001 have caused many restrictions on flight deck access. Pilots may observe operations in air traffic facilities under certain conditions, and are encouraged to do so. Using real air traffic controllers during simulator sessions has also proven beneficial to pilots and participating controllers. Aircraft dispatchers have functioned jointly with flight captains for years. They have been allowed, indeed required, to observe cockpit operations from the cockpit jumpseat as part of their initial and recurrent qualification under part 121. Some carriers have included day trips to their aircraft dispatchers' offices to provide the pilot insight into the other side of the joint function scheme. Those trips have commonly been part of the special training offered to first-time captains. Now, real-life aircraft dispatchers are increasingly being used in simulator sessions. The training experience gained by the pilot and the dispatcher is considered the logical extension of earlier training methods, providing interactivity where CRM principles are applied and discussed. Under certain conditions, maintenance personnel have had access to the cockpit jumpseat in air carrier operations under part 121; but that access has come under scrutiny because of security concerns following the attacks of 9/11. Training of first-time captains has often included day trips to a carrier's operations control center where a pilot and a maintenance supervisor can meet face to face and discuss issues of mutual interest in a thrumming, real-life setting. Dedicated CRM training courses for maintenance personnel have been operating since 1991. Even broader cross-pollination of CRM concepts has been considered, using other groups, such as passenger service agents, mid- and upper-level managers, and special crisis teams like hijack and bomb-threat teams.

Error Management. It is now understood that pilot errors cannot be entirely eliminated. It is important, therefore, that pilots develop appropriate error management skills and procedures. It is certainly desirable to prevent as many errors as possible, but since they cannot all be prevented, detection and recovery from errors should be addressed in training. Evaluation of pilots should also consider error management (error prevention, detection, and recovery). Evaluation should recognize that since not all errors can be prevented, it is important that errors be managed properly.

Advanced CRM. CRM performance requirements or procedures are being integrated into the special operating procedures (SOPs) of certain air carriers. Specific callouts, checks, and guidance have been included in normal checklists, the quick-reference handbook (QRH), abnormal/emergency procedures, manuals, and job aids. This integration captures CRM principles into explicit procedures used by flight crews.

Culture issues. While individuals and even teams of individuals may perform well under many conditions, they are subject to the influence of at least three cultures – the professional cultures of the individuals themselves, the cultures of their organizations, and the national cultures surrounding the individuals and their organizations. If not recognized and addressed, factors related to culture may degrade crew performance. Hence, effective CRM training must address culture issues, as appropriate in each training population.

When the airliner ditched in the Hudson, the coordination that occurred between pilots and flight attendants was an excellent example of teamwork. In the past, though, the results have not been so successful.

In 1989, a Foker F-28 crashed in Ontario, Canada during takeoff due to icing on the wings and killed 24 passengers. The flight attendants had seen wet snow building up on the wing but were trained to trust the judgment of the pilots and not question them. Additionally, one of the pilots had a reputation for not trusting the concerns of the flight attendants. Furthermore, airline policy discouraged the flight attendants from communicating with the crew. As a result, the flight attendants did not notify the pilots of the icing buildup.

In the same year, a 737 crashed at Kegworth, England after taking off from London Heathrow Airport. The crash killed 47 of the 129 passengers. The aircraft had experienced a vibration during climb out, which led the pilots to reduce power to the right engine. This power reduction reduced the vibration, and the pilots concluded that the right engine had simply malfunctioned. Accordingly, they made an announcement to the cabin that they were shutting down the right engine. In the meantime, the flight attendants could see flames coming out of the left engine. Even though the pilots had announced that they were shutting down the right engine, the flight attendants did not call to report that they had seen flames shooting out of the left engine. So a good engine was shut down while the left engine was on fire. During the approach to the airport, the engine burst into flames and ceased. In the end, the aircraft glided to a crash landing at the threshold of the runway. They almost made it.

The lack of teamwork between pilots and flight attendants has been identified as a major cause of accidents and can actually be traced to many different causes. First, pilots and flight attendants have almost opposite work schedules. During takeoff and landing, pilots are very busy with cockpit duties, and flight attendants are strapped in their seats. In cruise flight, the autopilot is engaged and pilot task loading decreases. Flight attendants, however, start their meal and beverage services just after level-off. These opposite work schedules can become an issue when things go wrong. Another issue is the observance of each other's duties on the aircraft. Deadheading pilots[23] can observe flight attendants working, but the flight attendants do not have the same opportunity. Since 9/11, there are restrictions that do not allow flight attendants in the cockpit during the critical cockpit work times (i.e., below 10000 feet). Another issue that could cause a breakdown in communication is cultural difference. Most pilots are male, and most flight attendants are female. Thinking processes differ between genders.

Recognizing that severe problems exist in crew coordination between pilots and flight attendants, the FAA issued AC 120-48 to generate recommendations for training and standard operating procedures that ensure more effective collaboration between pilots and flight attendants. The following is extracted from that advisory circular:

> In certain circumstances, it is important for flight crewmembers and flight attendants to act as one cohesive crew, even though

23 Deadheading means riding on a flight in the cabin to get to your next flight. At these times, a pilot rides as opposed to flying.

they are trained, scheduled, and generally regarded as two independent crews. When it is necessary to act as one crew, the activities of the cockpit and cabin should be coordinated. One of the prerequisites for crew coordination is effective communication between all crewmembers. In a 1986 survey of pilot safety representatives and flight attendants, only 37% of the flight attendants and 60% of the pilots said that they thought that communication between the cockpit and cabin is adequate. The key to improving coordination between flight crewmembers and flight attendants lies not only in improving communications between crewmembers but also in increasing flight crewmember awareness of flight attendant duties and concerns, and in increasing flight attendant awareness of flight crewmember duties and concerns. Seventeen percent of the flight attendants and 12% of the pilots surveyed said that their training did not cover each other's duties during emergencies; 51% of the flight attendants and 24% of the pilots said they did not cover each other's duties before takeoff and landing. During normal operations, it is important that each crewmember be familiar with the duties of the other crewmembers at every stage of the flight so that they can be sensitive to the other's level of workload. Such knowledge helps to avoid miscommunication, unrealistic expectations, and inappropriate requests of other crewmembers. During emergencies, each crewmember should know exactly what to expect from the other crewmembers so they can work together effectively.

Pilot Communication with Flight Attendants

Takeoff and Landing. It is vitally important that flight attendants are given adequate time to prepare the cabin and themselves for takeoff and landing, especially since most accidents occur during these critical phases of flight. Even when flight attendants are informed that takeoff is imminent, problems can arise that result in flight attendants not being properly seated for takeoff, particularly with unusually short taxi tines. Similar problems arise when flight attendants do not have adequate time to prepare the cabin for landing and take their jumpseats. The potential for problems is heightened when meal or beverage service is offered on very short flights (30 minutes or less). The most effective remedy for these problems is to have a flight attendant inform the captain, either by interphone or signal, that the cabin is secured for takeoff or landing. This procedure was regarded as important by 96% of the pilots and 91% of the flight attendants surveyed.

Turbulence. It is important that flight attendants receive timely notification of turbulence from the flightcrew. Flight crewmembers generally warn flight attendants of anticipated turbulence so that lack of such notice is not a common problem. However, it is one that can result in severe injury, particularly to flight attendants, since the majority of the serious injuries that occur as a result of turbulence are incurred by flight attendants. A member of the flight crew should inform the flight attendants of anticipated turbulence prior to the flight, since notification en-route may come too late to prevent injury. This is best accomplished by discussing en-route weather in a flight crewmember/flight attendant preflight briefing. While airlines consider this practice to be standard operating procedure, it is not always done. Only 56% of the flight attendants surveyed said that en-route weather is typically covered in a captain/flight attendant briefing. (However, 84% of the pilots surveyed reported covering it.) During the flight, flight attendants need to be informed of the immediacy and severity of unexpected turbulence so that they may determine whether to secure the cabin or to be seated immediately. On large turbojet airplanes, turbulence experienced in the flight deck may be much less than that experienced in the cabin. So, in some cases, flight attendants should advise the flightcrew about the severity of turbulence so that the seatbelt sign can be illuminated.

Emergencies. The most common examples of problems in communication during emergencies involve the flight crewmembers not informing the flight attendants of the nature of the emergency, the time available to prepare the cabin, and the necessary special instructions (e.g., to use only one side of the aircraft in the evacuation). This problem has arisen several times, despite instructions in flight manuals to relay such information to the flight attendants. The quality and timing of the information given to the flight attendants is extremely important in an emergency. Communications from the flightcrew should be clear, precise, and instructional. A vague description of the situation without specific instructions may be misinterpreted and result in valuable time being misspent. The timing of the information transfer is as important as the quality of the information. For example, when an aircraft will be landing without a functional nose gear and the captain decides to prepare for an emergency evacuation and to move passengers to the rear of the airplane, the flight attendants should be informed of the decision to move passengers at the same time that they are informed of the emergency so that they are aware of all the conditions before they select and instruct passengers to assist them in the evacuation. Also, in any emergency or unusual situation, it is important that the flight attendants be informed before the passengers, so that they have time to prepare.

Flight Attendant Communication with Pilots

Just as with cockpit-to-cabin communications, the timing and quality of the cabin-to-cockpit communications are critical. When flight attendants convey information to the flightcrew, the information should be timely and specific. The most common problems with cabin-to-cockpit communications can be divided into two categories: the failure of the flight attendants to convey important safety-related information to the flight crewmembers and inappropriate requests for information by flight attendants (i.e., breaking the "sterile cockpit" rule for reasons unrelated to safety). Both of these types of communication problems are related to the "sterile cockpit" issue. There are two major problems associated with flight attendant observance of sterile cockpit procedures; sterile cockpit time and sterile cockpit meaning.

Sterile Cockpit Time. It is difficult for the flight attendants to judge when sterile cockpit procedures are in effect. Flight attendants have no way of knowing when the aircraft is at 10000 feet unless they are told or signaled in some way. Some airlines have advocated the 10-minute rule, i.e., sterile cockpit procedures should be in effect for 10 minutes after takeoff and 10 minutes before landing. However, it is very difficult to estimate a time interval before an event.

Sterile Cockpit Meaning. Many flight attendants do not have a clear understanding of what "sterile cockpit" means. Eighty percent of the pilots and 86% of the flight attendants surveyed said this concept needs to be clarified for flight attendants. That is, flight attendants need to be given specific information about the purpose and meaning of the regulation and what type of information merits contacting flight crewmembers during the sterile period. There have been many instances of flight attendants going into the cockpit to request passenger information (e.g., on connections) or for other reasons not related to safety when sterile cockpit procedures were in effect. Such interruptions can distract flight crewmembers and have a detrimental effect on their performance. However, hesitancy or reluctance on the part of a flight attendant to contact the flight crewmembers with important safety information because of a misconception of the sterile cockpit rule is potentially even more serious than the unnecessary distraction caused by needless violations of the sterile cockpit. Flight attendants have failed to communicate to flight crewmembers important information concerning (e.g., fire in a galley trash container, a loud noise with vibration, and changes in cabin pressure) for fear of violating sterile cockpit procedures. Flight attendants should be aware that it is always important that they report unusual noises and abnormal situations to flight crewmembers as soon as possible and be specific in their report.

Emergency Procedures

Training is widely regarded as the most effective means of improving crew coordination. Statements in manuals, without the appropriate training, may not lead to the proper response in an emergency. Training for good crew coordination should include instructing flight crewmembers and flight attendants on each other's emergency procedures, codes, signals, and safety-related duties. In an emergency, it is imperative that each crewmember interpret emergency signals and codes in the same way. For example, code words or signals for hijacking or evacuation are useless unless each crewmember is aware of their meaning. Furthermore, emergency procedures for flight crewmembers and flight attendants should be compatible. For example, if flight attendants are taught that the second officer will occupy a cabin seat in preparation for a ditching in a certain aircraft, then flight crewmembers should be informed of this in their training. When manuals for flight crewmembers and flight attendants are written and revised independently, they should be cross-checked for consistency. Training administrators should ensure that the emergency procedures and other safety-related information presented to flight crewmembers is compatible with the information presented to the flight attendants. In any emergency, the flight attendants should know the nature of the emergency, the time available to prepare the cabin, what the bracing signal will be, and if there are any special instructions. When possible, the flightcrew should be ready to give the flight attendants this information in a timely manner. A well-orchestrated preparation for an emergency evacuation, or the handling of any other emergency, requires stressing the appropriate procedures in training for all crewmembers so that they act as a well-coordinated crew.

Normal Procedures

Coordination between flight crewmembers and flight attendants during normal operations also requires appropriate training. Crewmembers should be instructed on each other's safety-related duties and workload during preflight, takeoff, cruise, and landing. Such training helps to avoid miscommunication, unrealistic expectations and inappropriate requests of other crewmembers. Additionally, training should stress the types and quality of information that one crewmember expects from another. This is best accomplished by either having flight crewmembers and flight attendants in classes together or by having the same instructors teach flight crewmembers and flight attendants. The training material may also be covered by a flight attendant instructor participating in flight crewmember training and a representative of the flightcrew (e.g., instructor or check airman) participating in flight attendant training. A videotaped or slide presentation on each crewmember's duties can also be extremely effective, as well as cost effective, when presented by an instructor and discussed.

Cockpit resource management programs present an ideal opportunity to cover communication and crew coordination between flight crewmembers and flight attendants during flight training. However, training for flight deck/cabin communication should not be limited to captains, as cockpit resource management programs often are. First and second officers often handle all of the communications with the flight attendants. In fact, second officers usually act as the communication link between the flight deck and the cabin.

Flight attendants should receive special instruction regarding sterile cockpit procedures so that they neither naively violate them nor hesitate to communicate relevant information to the flightcrew. They should be given a clear, operational definition of the regulation and instructed as to when, and with what information, to contact the flightcrew. Flight attendants are typically instructed that they should not contact the flightcrew with information unless it is "safety-related". This directive alone leaves much room for interpretation. While it would be impossible to describe the kinds of information that should be relayed to the flightcrew, perhaps it would be helpful to give a few examples in training. The quality of the decisions (as to whether or not to contact the flightcrew) made by the flight attendants will be directly related to the information they received in training. The clearer the flight attendant's understanding of sterile cockpit procedures and flight operations, the better these decisions will be.

Practices and Procedures

There are many simple practices that can help to enhance the working relationship between flight attendants and flight crewmembers and which may be used to foster an atmosphere that is conducive to good communication. These practices include: respectful introductions, displays of common courtesy, announcements from the flight deck during delays to keep flight attendants and passengers informed, and the captain being supportive of flight attendants when problems arise in the cabin (e.g., a disorderly passenger). Perhaps the single most important procedure for setting the stage for good coordination between flight crewmembers and flight attendants on any flight is the flight deck/cabin (or captain/flight attendant) preflight briefing.

Cockpit/Cabin Preflight Briefing. A good flight deck/cabin preflight briefing gives the flight attendants the names of the flight crewmembers, the in-flight weather, the estimated flight time, and any unusual circumstances expected during the flight. Other topics can also be covered, such as flight deck entry procedures, a review of emergency communication procedures, details of the meal service, or any topic that any crewmember considers to be important. The briefing should allow crewmembers to solicit information from each other and to bring to the attention of the other crewmembers any information that they believe to be relevant.

Other Recommended Practices. Most of the recommended procedures are stated as company policy for many airlines. This indicates a need for these practices to be stressed during crewmember training as procedures to be followed on every flight. In addition to a flight deck/cabin preflight briefing, the following practices are highly recommended for optimizing crew coordination:

Warnings from the flight crewmembers to the flight attendants when the time between taxi and takeoff will be shorter than expected and when arrival time will be sooner than expected to give the flight attendants an indication of the time available to prepare the cabin for takeoff and landing;

Notification to the flight crewmembers from the flight attendants when all pre-takeoff and pre-landing duties have been completed and the cabin is secured;

Pre-takeoff and pre-landing signals or announcements from the flight crewmembers to allow sufficient time for the flight attendants to be seated;

Use of public address system to alert flight attendants and passengers of anticipated in-flight turbulence;

Notification to flight attendants when turbulence is severe enough to cease in-flight meal and beverage service and/or be seated with their restraints fastened, and when it is safe for them to resume their duties; and

Notification to flight attendants when "sterile cockpit" procedures are in effect. A good signal for this is an indicator light above the cockpit door or on the annunciator panel that has a duration as long as the sterile cockpit interval (as opposed to discrete tone or announcement that could be missed) and cannot be confused with another signal.

Many aviation accidents have been attributed to a breakdown in the aviation team. In an effort to reduce these errors, CRM training has been made mandatory at all airlines. Since errors inevitably arise, members of the flight crew must cross-check each other and be assertive when they see that something going wrong. And because we learn from our mistakes, advances in improving aviation teamwork have allowed the United States to create the safest aviation system in the world.

Chapter 12:

Pilot Workload and the Requirement for Automated Bird Warnings

Flying is a three-dimensional art that requires skill, situational awareness and the ability to think faster than the speed at which aircraft is moving. It takes many thousands of hours and years of flying to become a highly skilled pilot. To maintain proficiency, the skill must be practiced. Now, in order to understand the need for automation in the cockpit, it is important to first understand airline operations, including what it takes to become a major airline pilot, daily flying schedules, and the tasks a pilot must perform during the critical takeoff and approach phases of flight when most bird strikes occur. This chapter will begin by describing these operations and end with an assessment of the impact that various cockpit bird warning systems would have on airliner flight safety.

To become an airline pilot, there are two paths—one is through various types of civilian schools and training, and the other is through the military. A college degree is not required, however, most airlines look for it as evidence that you are trainable. If you choose the civil career path, you can attend colleges and universities that offer aviation degrees in parallel with flight training. There are also technical schools that offer flight training in less time than it takes to obtain a college degree. Another solution to education and flight training is to attend one of 600 flying schools. The advantage to this type of training is that you can do it on your own time.

It doesn't really matter how or where you get your education. What counts most is that you pass the written and flight evaluations. In order to pilot an aircraft in the United States, a pilot must possess a valid airman certificate. The first certificate is a Student Pilot Certificate. To qualify for a student pilot certificate, an individual must be 16 years old; read, speak and understand English; and be able to pass a third-class medical examination. At that point, an FAA-authorized aviation medical examiner will issue you a combined medical certificate and student pilot certificate.

To qualify for a third-class medical certificate, pilots must have distant vision (20/40 or better in each eye separately, with or without correction); near vision (20/40 or better in each eye separately, with or without correction, as measured at a distance of 16 inches); color vision (demonstrate the ability to perceive the colors necessary for the safe performance of airman duties); hearing (demonstrate the ability to hear an average conversational voice in a quiet room, using both ears, at a distance of six feet, with their back turned to the examiner, or pass an approved audiometric test); ear, nose, and throat (exhibit no ear disease or condition manifested by, or that may reasonably be expected to be manifested by, vertigo or a disturbance of speech or equilibrium); blood pressure (under 155/95); mental status (no diagnosis of psychosis, bipolar disorder or severe personality disorders) and substance dependence (no dependence on alcohol or any pharmacological substance in the previous two years).

You must have a Student Pilot's License to solo an aircraft. Before you solo, though, you must pass a written test and receive training from a certified instructor. This certificate allows a student pilot to operate an aircraft with an instructor pilot, or after several hours of instruction and under certain limitations, the student pilot can fly alone. The limitations include flying solo in controlled airspace, within 25 NM of specified airports, in a specific make and model aircraft. Student pilots cannot carry passengers and cannot be paid for flying. The weather must be clear (i.e., 3 miles of visibility during the day and 5 miles at night)

The next certificate in the progression to becoming an airline pilot is called a Private Pilot's License. The general eligibility requirements are that you be 17 years old; able to read, speak and understand English; receive a logbook endorsement from an instructor; pass a written exam and have 40 hours of flight time (normally, it takes the average person 60-70 hours to master flying), which is broken down into a minimum of 20 hours of training from an instructor and 10 hours of solo flight. Other flying can be a combination or solo flight. You must demonstrate aeronautical knowledge in the following areas: Federal Aviation Regulations (FARs) that relate to private pilot privileges, limitations, and flight operations; accident reporting requirements of the National Transportation Safety Board; use of the applicable portions of the 'Aeronautical Information Manual' and FAA advisory circulars; basic aerodynamics and the principles of flight; use of performance charts; significance and effects of exceeding aircraft performance limitations; principles and functions of aircraft systems; night and high-altitude operations; descriptions of and procedures for operating within the National Airspace System; use of aeronautical charts for VFR navigation using pilotage, dead reckoning, and navigation systems; radio communication procedures; recognition of critical weather situations from the ground and in flight, windshear avoidance, and the procurement and use of aeronautical weather reports and forecasts; safe and efficient operation of aircraft, including collision avoidance, and recognition and avoidance of wake turbulence; effects of density altitude on takeoff and climb performance; weight and balance computations; principles of aerodynamics, power plants, and aircraft systems; stall awareness, spin entry, spins, and spin recovery techniques; aeronautical decision making and judgment; and preflight action that includes how to obtain information on runway lengths at airports of intended use, data on takeoff and landing distances, weather reports and forecasts, and fuel requirements; how to plan for alternatives if the planned flight cannot be completed or delays are encountered;

and maneuvers, procedures, and emergency operations appropriate to the aircraft.

In addition to having academic knowledge of the areas listed above, you must also demonstrate flying proficiency in the following areas: preflight preparation; preflight procedures; airport and seaplane base operations; takeoffs, landings, and go-arounds; performance maneuvers; ground reference maneuvers; navigation; slow flight and stalls; basic instrument maneuvers; emergency operations; night operations and postflight procedures. After you have logged your hours and passed your written ground school test, you'll need to pass a check-ride with an FAA examiner. The examiner will ask you to plan a flight, ask you questions, and require you to fly certain maneuvers. If everything goes well, the examiner will issue you a private pilot's certificate.

A Private Pilot's License can be obtained for the following aircraft categories: airplane, balloon (gas), balloon (hot air), glider, gyroplane, helicopter, lighter-than-air (airship), rotorcraft gyroplane and rotorcraft helicopter. After you pass the flight test, you are required to maintain a certain proficiency in order to maintain currency. Before carrying passengers, you must have made at least 3 takeoffs and 3 landings within the previous 90 days; and in order to maintain night currency, you must have made at least 3 full-stop landings within the previous 90 days. If you lose your currency, you must fly with an instructor.

To make matters more confusing, a private pilot can receive endorsements (ratings) that include: instrument rating, multi-engine, piston/turbine, tail wheel, retractable landing gear, float-plane, aerobatics, spins, formation flying and crop dusting. To become an airline pilot, you must have the instrument and multi-engine endorsements. To obtain the instrument endorsement, you must first pass an exam on instrument flight rules (IFR), IFR navigation, use of IFR charts, weather, decision making and crew resource management. In addition to the academic requirements, you must have at least 50 hours of cross-country experience (i.e., a flight of more than 50 miles from the airport) and 40 hours of actual or simulated instrument time. After you meet all of these requirements, you can take a flight test with an FAA examiner. Once you pass the test, you are required to maintain proficiency to maintain currency. You must log 6 instrument approaches every 6 months. If you fail to meet this requirement, you must pass an instrument proficiency flight test (i.e., check ride) administered by a certified flight instructor with an instrument endorsement (CFII). The second endorsement you must have in order to become an airline pilot is a multi-engine rating. There is no requirement for a separate written exam, but you must take a flight test with an FAA-certified flight examiner.

The next stop on your road to becoming a major airline pilot is the Commercial Pilot's License. The basic requirements for a commercial license are: read, speak, write, and understand English; be 18 years old; hold at least a Private Pilot's License; receive and log the appropriate ground and flight training for the commercial license; have 250 hours total flight time; have 100 hours flight time as pilot in command; have 50 hours of cross-country flight time as pilot in command; pass the FAA commercial pilot written exam and pass the commercial pilot oral and practical exam. A pilot must also pass a second-class medical exam with the following requirements: distant vision of 20/20 or better in each eye separately, with or without correction; and intermediate vision of 20/40 or better in each eye separately, with or without correction, at age 50 and over. Once you have the commercial license, you can be paid for flying. Once you reach this level of flying, you have a much greater knowledge of aircraft systems and the ability to make better decisions in the air when things don't go quite right. This ability is called airmanship.

The final stop on your way to becoming a major airline pilot is to obtain the Airline Transport Pilot (ATP) License. The requirements are that you must be at least 23 old; be able to read, speak, write, and understand English; be of good moral character; and meet at least one of the following: hold at least a commercial pilot certificate and an instrument rating; meet the military experience requirements (listed in FAR 61.73) to qualify for a commercial pilot certificate and an instrument rating; hold either a foreign ATP or a foreign commercial pilot license and an instrument rating, without limitations, issued by a member nation of the International Civil Aviation Organization (ICAO); receive and log ground training from an authorized instructor covering the following areas: applicable FARS that relate to airline transport pilot privileges, limitations, and flight operations; meteorology, including knowledge of and effects of fronts, frontal characteristics, cloud formations, icing, and upper-air data; general systems of weather and NOTAM collection, dissemination, interpretation, and use; interpretation and use of weather charts, maps, forecasts, sequence reports, abbreviations, and symbols; National Weather Service functions as they pertain to operations in the National Airspace System; windshear and microburst awareness, identification, and avoidance; principles of air navigation under instrument meteorological conditions in the National Airspace System; air traffic control procedures and pilot responsibilities as they relate to en-route operations, terminal area and radar operations, and instrument departure and approach procedures; aircraft loading, weight and balance, use of charts, graphs, tables, formulas, and computations, and their effect on aircraft performance; aerodynamics relating to an aircraft's flight characteristics and performance in normal and abnormal flight regimes; human factors; aeronautical decision making and judgment; crew resource management to include crew communication and coordination.

You must pass a written test covering the areas above and accumulate 1,500 hours of flying time, which is broken down into at least 500 hours of cross-country time, 100 hours of night time, 75 hours of instrument time, and 250 hours as pilot in command. A final requirement is to pass a first-class physical, which basically has the same requirements as a second-class physical except that you must have an electrocardiogram on an annual basis once you reach 40 years of age. For pilots under 40 years of age, first-class medical certificates expire on the last day of the month they were issued, one year from the date of issue. Once you meet all the requirements above, you need to take a flight test with an FAA examiner that covers the following areas: preflight preparation, preflight procedures, takeoff and departure phase, in-flight maneuvers, instrument procedures, landings and approaches to landings, normal and abnormal procedures, emergency procedures and postflight procedures.

Even after accomplishing all of these steps, though, the major airlines still will not hire you. The average number of hours for hiring pilots with civil experience is 4,000 to 6,000 hours; and it can take 8 to10 years to accumulate this amount of flying time. What most pilots do is work for a smaller regional airline for a while to accumulate these hours and then try to move up, assuming the economy is not in a slump, which normally requires the airlines to layoff or furlough pilots. Another path to becoming an airline pilot is through the military. After a year of pilot training, pilots must agree to a 10-year service commitment. You must be a college graduate and normally receive a commission through the academies, ROTC or other training schools. After leaving the military, you normally take a few exams to obtain the civilian ratings listed above, and then you go through the same application process as civilian pilots.

So, assuming that you now have accumulated the necessary hours and a major airline has an opening, you apply for a job. When I was interviewed, the hiring department told me that only 3 out of 10 would be hired. Also, I found out that my airline had a database of over 10,000 applicants. So, there's a bit of luck involved in being hired. In addition to the training and licenses mentioned above, an airline will normally require you to pass both an aptitude and psychological test. This is because pilots need to make quick decisions. The licenses described above remain valid as long as you can pass a physical and eye examination and periodic flight tests. These flight tests are mandated by both FAA regulations and airline company requirements.

If you pass the interview and are selected for hiring, you still have to go though yet another training program to be certified to fly a specific type of aircraft (called a 'type rating'). The airline has to train you using their FAA-approved training program. Furthermore, every airline has different procedures to follow while flying. This initial training will take 8 to 10 weeks to accomplish. In academics, you become an expert on the aircraft systems. After academics, you will enter the simulator training phase, where you will practice the normal procedures that are followed every time you fly. But the most challenging part is flying with simulated emergencies. Everything and anything that could go wrong is thrown at you during this training so that you're ready when something happens in the real aircraft. At the end of this training period, you will receive an oral examination and a check ride in the simulator.

Now, you are finally ready to fly a real jet! Normally, you fly 25 hours with an instructor to gain initial operating experience (IOE). After that, you're released as a regular crewmember to fly the line, and you'll work 80 to 100 hours per month. In order to control fatigue, the FAA regulates the maximum number of hours a pilot can fly. The limits are 30 hours in 7 days, 100 hours in 30 days and 1,000 hours in a 12-month period. Most pilots fly an average of 70 to 80 hours per month, but you will start out as a copilot or first officer and fly in this position for many years. The airlines have a seniority-based system, which means that your opportunity to move into the left seat as a Captain will take a while. Even after coming this far, you still are required to return to the simulator every 6 to 12 months to receive recurrent training and testing in the simulator to demonstrate proficiency.

A pilot's flying schedule is awarded each month based on seniority. Normally, new pilots are placed on a reserve status – they fly when the airline calls. This happens when someone gets sick or can't get to work because of a weather or mechanical problem. You are normally assigned to an on-call status with a 12-hour window, during which you must be within 2 hours of the airport and have a bag packed and ready to go at a moment's notice. You never know where you'll be asked to go or for how long you will be gone, so you have to pack for all contingencies. After several days in a row of on-call status, pilots are given several days off. This status is normally assigned to the newest, most junior pilots and is not much fun because you don't have much control of your life.

Once you are with the airline for a few years, you can move up to a regular schedule. Pilots input their requested schedules and destinations once per month and are awarded a line of flying time. A defined schedule means you know the days you'll fly and the days you'll have off. A day in the life of an airline pilot can vary based on a number of factors. For domestic (US only) operations, only 2 pilots are required, and the days may be as long as 16 hours, depending on the circumstances. You may be scheduled for a normal flying day of 8 hours, but a maintenance or weather delay may cause the day to stretch out to a maximum of 16 hours. International flying has different rules. If the flight goes over 8 hours but remains below 12 hours, then the crew must be augmented with an additional pilot. Pilots take turns flying and resting in an assigned seat in the cabin. If the flight goes over 12 hours, then there must be a total of 4 pilots. Pilots often experience jet lag, which causes fatigue when flying though time zones. This is particularly true for international flying. In order to combat the fatigue, the federal regulations require a pilot to have 8 hours off in the United States and 24 hours off internationally.

The NTSB has determined that fatigue has been a factor in many aircraft accidents and has recommended that the FAA solve it. The FAA has recognized that pilot fatigue is a real safety issue, but they believe that the issue is too complex. The airlines, however, don't want flying time to be regulated because it will limit a pilot's usefulness and reduce revenues. In 1995, the FAA published rules to limit flight and duty times, but it has not enforced them. As far back as 1989, the NTSB recommended setting working hour limits on flight crews, air traffic controllers and aviation mechanics to prevent fatigue. What's worse, current pilot flight time and rest rules are over 60 years old. ALPA has worked for decades to change the rules. In 2009, when pilot fatigue was identified as a safety priority for ALPA, the FAA finally announced plans to develop new standards for pilot flight time and modernize the regulations.

A typical flight involves a high degree of planning coordination, skill and quick thinking, all of which are learned over a long period of time. Before you arrive at the aircraft, you normally meet as a crew to carefully plan the flight. You plan your altitude, speed and route of flight using designated points in the sky, as if it were a sort of airborne highway. You check the weather at the departure, en-route and destination points. You discuss the departure and destination airport procedures and special information related to another country you may be flying to. Then, once the planning is complete, an IFR flight plan is filed.

After arrival at the aircraft, a very thorough preflight is conducted. A pilot checks the exterior and interior of the aircraft to ensure that everything looks and feels normal. An airliner has many complex and redundant systems, and it is important to make sure all of these systems are operating correctly before takeoff.

Special procedures are developed to test the aircraft systems and instruments on the ground before takeoff. Once everything checks out, the final paperwork is received. This usually consists of a fuel slip, an updated weather forecast and weight and balance numbers. The fuel slip must be reviewed to make sure there is enough gas on board for the flight. The weight and balance is important because it lists the final number of passengers, cargo, aircraft weight and fuel. Aircraft weights can range from 300,000 to 1,000,000 pounds. These numbers are used to compute various speeds to ensure the aircraft can be safely maneuvered. Other factors that affect the computation of these speeds are airport elevation, temperature, winds and runway condition.

The first speed is called V_1 and is associated with an aircraft abort. An aircraft could be struck by birds during takeoff and certain decisions need to be made quickly. The V_1 speed is a decision speed and is used to decide whether to abort or takeoff. If you are below the V_1 speed, then the decision is normally made to abort. If you are above V_1, then the decision is normally made to takeoff. However, if you are above V_1 and abort, you will most likely go off the end of the runway and have a very bad day. The second speed you compute is called V_r. This is the speed at which you pull back on the controls and rotate the aircraft into the air. The final speed is called V_2 and is associated with an engine failure. V_2 is the minimum speed you want to fly if you lose one of your engines on takeoff. All of these speeds will change depending on the aircraft's weight at that particular moment, so they are important to compute for each flight.

Flying is not a physical profession, but a lot of thinking is required to execute a safe and successful flight. A pilot must be ready to react to any abnormal event, especially during takeoff and landing. These are the most critical phases of flight because you are moving fast and close to the ground, so there isn't much margin for error. There are many things a pilot must be thinking about simultaneously in order to accomplish a successful takeoff. You must consider which way the winds are blowing so that you don't drift off the runway after liftoff. You need to know the initial altitude, airspeed and heading you will fly. You need to be thinking about how you will navigate the departure procedure. You also need to talk on the radio to the tower controller to request and acknowledge takeoff clearance. You need to make various configuration changes to the aircraft. The gear has to be raised and the flaps need to come up according to a speed schedule. You need to fly the aircraft precisely in three dimensions while controlling your airspeed, altitude and heading. You need to look outside to make sure you don't bump into another aircraft. You need to keep track of where you are and think ahead as to where you are going. You also need to accomplish certain cockpit checklists to make sure the aircraft is "cleaned up" (i.e., gear and flaps up) and ready to fly. Another issue to add to the complexity of a takeoff is bad weather. Pilots are trained to trust their instruments when they can't see the ground. You must be able to fly the aircraft precisely using only the displays in the cockpit. Instrument flying in bad weather is another complexity you can add to the list of items that need to be thought through before takeoff.

You also need to be prepared to abort should something like a bird strike occur below V_1. You need to be prepared for the most difficult emergency of losing an engine on takeoff. In a simulator, pilots are trained once or twice a year on the loss of an engine on takeoff. I can tell you from my personal experience that this is one of the most difficult maneuvers a pilot can fly, and it requires skill and precision. Various situations can cause an engine loss, including engine failure, engine seizure, bird strike, compressor stall and fire. After an airborne engine failure, all of your attention is focused on controlling the aircraft. You work with the other pilot as a crew to execute the checklist and make the aircraft safe to fly. In most cases, you also coordinate with the controller to setup a return to the airport for a landing. Airliners are designed to fly on one engine but must be handled with extreme caution and care.

The items above represent a comprehensive list of tasks that are on a pilot's mind during takeoff. Before I move on to the other phases of flight, I want to discuss the requirement for automation for a bird strike warning system. Later chapters will discuss, in greater detail, three methods for delivering bird target data to the cockpit: text, graphic and real-time display. From the sheer abundance of tasks listed above, you can clearly understand that a text or graphic display of bird targets will be of little use to a pilot once he is in the air.

Only an automated warning system will be usable to a pilot in the air during takeoff. Another implementation for bird strike warning could be used before takeoff. A pilot could receive the bird target data while he is on the ground. This would allow him to plan the takeoff to avoid the birds. If a pilot could see the bird targets overlaid on a navigation display, he could then decide whether the birds will be a conflict. He could simply ask the controller for a delay to allow the birds to fly through the intended flight path. On the ground, both text and graphic displays could be used for warning, although non-real-time information may have limited use. Once the takeoff is complete, the aircraft commences a climb to a final altitude. Since 92% of bird strikes occur below 3000 feet, the chances of running into a bird significantly decrease as altitude increases.

The next critical phase of flight occurs on approach and landing. During this phase, the aircraft will descend into the critical bird strike area of below 3000 feet. The planning for this phase occurs in cruise flight at high altitude before the descent even begins. The pilots check the weather and airport special items (called 'Notices to Airman' or NOTAMs). The procedures for flying around the airport and the navigation routes are examined. During the approach phase, you must think ahead as to what actions you will be performing – you will be descending, setting up your approach on the aircraft instruments, executing cockpit checklists, looking out for traffic, listening and executing the controller's instructions, slowing down, lowering the flaps, lowering the landing gear and flying the aircraft on a precise glide path to land. So, once again, you become task-saturated during this phase of flight.

Consider the use of a bird warning system in the cockpit before the descent is started several hundred miles from the airport. Bird target positions could be sent using a text, graphical or real-time service. A text-based service would need to take bird target locations and convert them into a usable format, but the problem with this approach is that the data is non-real-time. When the aircraft finally gets closer to the airport, the data will be old and the position of the bird targets will not be current. A graphical picture of the bird target could also be sent to the cockpit, but again, this would need to occur well in advance of the landing. If the picture is static, you would have to check it far from the runway and would again have the problem of old data being present by the time you land. If the picture is updated, much like weather displays, then a more real-time update could be received closer to the airport. However, the closer the aircraft gets to the airport, the higher the task loading is for the pilot. This could make a graphic display less useful as a pilot may not have time to reference a separate display or picture. The final method for presenting bird targets is on an existing cockpit display. This will give the pilot real-time, actionable data.

The aircraft route of flight is presented on a cockpit display and is the primary means of navigation. If bird targets could be overlaid on the same display, a pilot could determine whether a potential for collision exists. If that scenario develops, a pilot could take several courses of action to avoid a collision, one of which could be to ask the controller for a 360-degree turn. This would allow time for the birds to fly through the aircraft's projected flight path. Once the birds are clear and the 360-degree turn is complete, the aircraft could continue the approach to a landing. A second maneuver that a pilot could fly is called a go-around. In high-density airspace, a 360-degree turn will slow things down and have a ripple effect that delays the traffic behind. To avoid delaying other flights, a pilot could simply push the power up, retract the landing gear and flaps, and come around for another approach. This would of course delay the flight by 10 to 15 minutes and consume more fuel, but it would avoid the damage caused by a collision from a bird.

A third operational solution that could be accomplished is to make a series of turns. This maneuver would have to be accomplished before the final approach phase when the aircraft is lined up with the runway. The turns would cause a delay and would allow time for the birds to fly through the aircraft's projected flight path. When the birds are clear of the flight path, the aircraft would stop the turns and continue for a landing. If the bird condition at the airport becomes severe, a landing may be inadvisable. In this case, a pilot could ask to be directed to a holding pattern away from the airport. The aircraft would continue to hold until the bird condition clears up. After that, the controller would direct the pilot to leave holding and give him directions to the airport.

Real-time displays can also be incorporated into the control tower. As a matter of fact, the technology exists to send bird targets to both the cockpit and controller simultaneously. Adding the tower controller as an observer of bird targets and director for aircraft maneuvers could prevent a bird collision. There is actually an FAA regulation that directs controllers to notify pilots when birds are present. There have been bird strike accidents where, even though the controller saw the birds and advised the pilot, the pilot still made the decision to take off and, as a consequence, hit the birds. When interviewed about the accident, both the pilot and controller basically said "what was I supposed to do about it?"

Controllers know the flight path that an aircraft will follow during departure, so if the bird targets are overlaid on the controller's display, then he could predict a potential collision. The controller could advise the pilot of the location of the birds; however, the pilot will always make the final decision. On takeoff, a controller could either recommend a delay or recommend/direct the pilot to fly a different heading. Once the aircraft is airborne, the controller could actually direct the pilot to fly a series of headings to avoid the birds. The same operational procedures could be applied to approach and landing. If the controller sees a conflict on approach, he could again direct the pilot to fly a series of headings to avoid the birds.

The operational procedures listed above are hypothetical and were created by using my technical and operational experience. The final procedure will most likely be developed by RTCA, which is a private non-profit company that forms committees and provides recommendations for technical requirements and operational procedures for the FAA. RTCA has an air traffic management committee that tackles problems to improve safety, capacity and/or efficiency of the US air transportation system. When new equipment is added to the airports or control towers, RTCA is normally involved in discussing all aspects associated with the new item. They consider development costs, acquisition, facility and equipment modification, training, operation and maintenance and removal from service or all system life cycle issues. The most frequent items requested are new Minimum Operational Performance Standards (MOPS) or appropriate technical guidance documents. The RTCA forms a special committee to develop these documents, which then become the basis for certification.

Essentially, all RTCA products are developed by special committees staffed by volunteers. As with all Federal Advisory Committee activities, special committee meetings are publicly announced and open to participation by anyone with an interest in the topic under consideration. During special committee meetings, volunteers from government and industry explore the operational and technical ramifications of the selected topic and develop consensus-based recommendations. These recommendations are then presented to the RTCA Program Management Committee, which either approves the special committee report or directs additional work. Approved recommendations are published and made available for sale to members and to the public. Easy access to updates on committee activities and related subjects is available on the RTCA website and in the Digest, which is published every two months.

Through the years, RTCA has received several awards for its service to the aviation community. The organization was awarded the 1949 Collier Trophy for "A guide plan for the development of a system of air navigation and traffic control for safe and unlimited aircraft operations under all weather conditions." Additionally, in 1994, the FAA named RTCA, Inc. as the US recipient of the ICAO 50th anniversary Medal of Honour. This unique recognition identified RTCA as the single most important US contributor organization to the advancement and support of civil aviation since the Chicago Convention created the ICAO in 1944.

The final chapters of this book describe existing technologies that could be integrated to create text, graphic or real-time displays of bird targets in the cockpit. Even though the technology currently exists to accomplish this goal, current levels of interest and funding have been insufficient. For this reason, public awareness is essential. The increasing number of mid-air collisions in the 1980s made the public more aware and eventually led to regulatory action to require a technical solution. This solution took over 6 years for the airlines to implement despite its obviousness. The airliner in the Hudson has made the public more aware, but this awareness has already started to diminish. It won't be until there is blood on the runway before any action will be taken. But what form will that action take? ALPA is already complaining that the solution has taken too long. Hopefully, this time, we will do more than scare birds away or simply see birds on a radar. A funding source for a real-time cockpit display of bird targets should be identified. These funds could come from the government agencies that are already involved with the problem. Private funding is difficult to obtain because the payoff seems uncertain. One way or another, though, a prototype should be developed and demonstrated. As soon as possible, pilots need to be trying out the display during normal operations. Surveys could be used to collect pilot inputs and make modifications to the display system. In parallel with prototype development, a committee needs to create a list of operational procedures that a pilot should follow in event of a potential collision. Once these procedures are established, both pilots and controllers should be trained. Controllers need to know what pilots will do in order to avoid a potential collision. Pilots will need to be trained and tested, both academically and in the simulator, on execution of proper avoidance procedures. This process will have to be thorough, so it will have to be time-consuming; and the sooner we get this process going, the sooner we can start seeing results.

Chapter 13:

Cockpit Data Link Implementations for Bird Warnings

So far, I have explained how radar can be used as a sensor to detect and track bird targets. The next component of the collision avoidance system for birds is the data link from the land-based radar to the cockpit. Data links have been used in airline operations for over 30 years. Existing data links to the cockpit include addressable point-to-point and the more recent broadcast data links associated with the Next Generation (NextGen) air transportation system. This chapter describes the data links and their possible integration into a bird strike collision avoidance system.

The recommended data link implementations are broken down into the categories of real/near-real-time and non-real-time. Currently, airline cockpit displays display the aircraft's present position and future flight path. The real/near-real-time implementations would provide bird target data directly to the cockpit so that the pilot could see bird targets moving on the overlaid aircraft flight path. The non-real-time implementation would send a bird snapshot image from the bird radar display to the cockpit either automatically or on demand. An image processor would need to be added to the radar data processor to time tag, capture and archive images of the bird target display. The non-real-time image could be sent on demand to the pilot just before takeoff or in the descent before the approach phase of flight. The final portion of this chapter describes some other advanced concepts that use the Internet and in-flight pico cell phone technology.

Aircraft Communications Addressing and Reporting System (ACARS) is a digital data link system for transmission of short, relatively simple messages between aircraft and ground stations via radio or satellite. The airlines, in an effort to reduce crew workload and improve data integrity, introduced the ACARS system in the late 1980s. This data link is designed to send and receive digital messages from the ground by means of existing VHF radios. With over 1000 stations in over 160 countries, ACARS is the world's largest air-to-ground VHF data network.

One of the initial applications for ACARS was to automatically detect and report changes to the major flight phases (i.e., Out of the gate, Off the ground, On the ground and Into the Gate), which are referred to in the industry as OOOI. During the late 1980s and early 1990s, a data link interface between ACARS and Flight Management Systems (FMS) was introduced, enabling flight plans and weather information to be sent from the ground to ACARS, which would then be forwarded to the FMS. This feature gave the airline the capability to update FMSs while in flight, and allowed the flight crew to evaluate new weather conditions or alternate flight plans.

In the 1990s, the ACARS system was improved to allow engine, aircraft, and operational performance conditions to be sent to the company via satellite in the ACARS format. This reduced the need for airline personnel to go to the aircraft to off-load the data. Abnormal flight conditions are automatically sent in real time. This capability enabled airlines to better monitor their engine performance and identify and plan repair and maintenance activities. In addition to transmitting engine data, ACARS was upgraded to transmit maintenance data. This enabled airline maintenance personnel to receive real-time data associated with maintenance faults on the aircraft. Airline maintenance personnel could now start planning repair and maintenance activities while the aircraft was still in flight.

All of the processing described above is performed automatically by ACARS and the associated avionics systems, with no action performed by the flight crew. As part of the growth of the ACARS functionality, ACARS also interfaces directly with a control display unit (CDU) located in the cockpit. This CDU, often referred to as an MCDU or MIDU, provides the flight crew with the ability to send and receive messages similar to today's email. To facilitate this communication, the airlines, in partnership with their ACARS vendor, define MCDU screens that are presented to the flight crew and enable them to perform specific functions. This feature provides the flight crew flexibility in the types of information requested from the ground and the types of reports sent to the ground.

Airlines began adding messages to support new applications (i.e., weather, winds, clearances, connecting flights, etc.), and the ACARS systems became customized to support airline-unique applications and unique ground computer requirements. This resulted in each airline having their own unique ACARS application operating on their aircraft. In fact, some airlines have more than 75 MCDU screens for their flight crews, where other airlines may have only a dozen different screens. In addition, since each airline's ground computers were different, the contents and formats of the messages sent by ACARS were different as well.

The majority of ACARS messages are typically only 100 to 200 characters in length. Such messages are made up of a one-block transmission from (or to) the aircraft. The text message block is constrained to be no more that 220 characters within the body of the message. The bandwidth of this data link is only 2.4 kilobits per second. Graphical weather can be obtained by using compression algorithms to reduce a graphic file to a character string that is uplinked through ACARS.

The same compression algorithm could be used to provide a bird snapshot from a bird radar display. The image could be provided at the request of the pilot.

Digital Automatic Terminal Information System (D-ATIS)

Regular voice ATIS is a continuously transmitted VHF voice message that a pilot listens to, either before takeoff or before approach and landing, in order to obtain essential information about the airport. The automation of this message reduces frequency congestion and relieves the controller of the need to repeat this information to every aircraft. The tower controller either makes an audible recording and then enables it for playback or enters the information into an automated system that uses a speech synthesizer for playback.

Many high-capacity airports employ the use of D-ATIS, which allows the controller to enter the airport information into a database and the pilot to send a request through the VHF ACARS data link to retrieve the text-based message. The following information is contained in an ATIS message: airport name; ATIS phonetic alphabet code (e.g., information "Bravo"); UTC time (Zulu time[24]); cloud ceiling, measured or estimated; visibility in miles and/or fractions of miles; temperature in degrees Fahrenheit; dew point in degrees Fahrenheit; wind direction and speed; altimeter setting instrument approach procedures in use; runway(s) in use for arrivals; runways in use for departures; pertinent NOTAMs or weather advisories; braking action reports if appropriate; low-level wind-shear advisories if appropriate; remarks or other information; and controller identification information.

The following is an example of a D-ATIS message:

KEWR ATIS INFO O 1751Z. 16010KT 9SM FEW018 BKN080 23/18 A3008 (THREE ZERO ZERO EIGHT). ILS RWY 22L APCH IN USE. DEPARTING RWY 22R. RWY 11/29 CLSD. NORTH 4 HUNDRED AND FIFTY FEET OF RY 22R CLSD. RY 22R ALD 9 THOUSAND 5 HUNDRED AND FIFTY. RY 22R DEPARTURES AUTHORIZED FROM INTERSECTION Y, AVAILABLE DEPARTURE DISTANCE 9 THOUSAND 5 HUNDRED AND FIFTY. READBACK ALL RUNWAY HOLD SHORT

---

24 Zulu time is Greenwich Mean Time, the standard from which all time in the world is referenced.

INSTRUCTIONS. **USE CAUTION FOR BIRDS IN THE VICINITY OF EWR.** ...ADVS YOU HAVE INFO O.57DE

A pilot would read the abbreviated message as: Newark arrival information oscar, time 1715 Zulu. Winds 160 degrees at 10 knots, visibility 9 statute miles, few clouds at 1800 feet, broken clouds at 8000 feet, temperature 23 degrees, dew point 18 degrees, altimeter 3018. The landing runway is 22 Left, and the departing runway is 22 Right. Runway 11/29 is closed. The rest of the transmission deals with takeoff, landing and taxi instructions, followed by some advisories. Note the bold-faced text describing the bird warning. This is currently the only warning a pilot will receive of birds being a threat at the airport. "Use caution for birds in the vicinity of the airport" or "moderate bird activity in the vicinity of the airport", etc. During the NTSB hearings, the pilot of the aircraft that ditched in the Hudson was asked about the ATIS warning system, and his answer was that "these warnings are general in nature."

Since this is a text-based data link, it would be a non-real-time data link for implementation in an aircraft bird strike warning system. Pilots need to know exactly where the birds are, not just that they are present. After the bird targets are processed by the radar, additional information on bird target location relative to takeoff and landing runways could be added to the D-ATIS transmission. The bird latitude, longitude and altitude could be computed, and the center of the landing runway could be used as a reference. An example D-ATIS transmission could be:

USE CAUTION FOR MODERATE BIRD ACTIVITY AT 3 NM, 1300 FEET AND 30 DEGREES LEFT OF RUNWAY HEADING

While this transmission is not real-time, it does provide pilots with more information about where the birds are and whether they will be a threat. A pilot does not have a display, but this text message will allow a pilot to visualize where the birds are relative to the runway. The pilot knows how far he is from the airport and can compare the bird range with the aircraft range to landing. The aircraft altitude is also known and can be compared to the bird target altitude to determine whether a hazard exists. The data would have to updated in real time in post-radar data processing. The drawback to this approach is that the information is only as good as when it was last received. If a pilot requests the data 20 minutes before landing, then it will be very old by the time his landing occurs. Therefore, this approach would be more useful for the takeoff phase when an aircraft is still on the ground and trying to decide whether to accomplish the takeoff or delay it for bird hazards.

VHF Digital Link Mode 2 (VDL-2)

VHF Digital Link Mode 2 or VDL-2 can be used to display graphical weather, electronic charts and engine/aircraft health monitoring programs and is used to enhance flight efficiency and safety. VDL-2 delivers data at 31.5 kbps, over 10 times faster than the rate used by ACARS. Aside from its much greater bandwidth, VDL-2 offers significant advantages over ACARS. Its internationally approved standards-based architecture provides tremendous user flexibility, including complete freedom of choice in aircraft and ground display systems, avionics, and applications. Because it offers a common infrastructure that can be shared by the entire aviation industry, its cost can be distributed over a large pool of users. ACARS ground stations are being replaced with integrated ground stations that support both ACARS and VDL-2. Eventually, VDL-2 will replace ACARS.

Bendix/King, a company owned by Rockwell Collins, has a state-of-the-art VDL-2 data link system to bring high-resolution NEXRAD (described in Chapter 5) and METAR graphical weather data to the cockpit. METAR is the most popular format in the world for transmission of weather data. The hardware required includes a KMD 550 receiver, KMD 850 Multi-Function Display and KDR 510 VDL-2 receiver. Once the equipment is installed, the user must subscribe to the Wingman™ service. This service is provided through 220 VDL-2 ground-based transmitters throughout the country. Compression algorithms are used to reduce a graphic file to a character string that is uplinked through a VDL-2 data link.

The same compression algorithm could be used to provide a bird snapshot from the bird radar display. The image could be provided at the request of the pilot.

NextGen

The Next Generation (NextGen) Air Transportation System contains three different cockpit data links: 1090 Extended Squitter (1090ES), Universal Access Transmitter (UAT), and VDL Mode 4. These three data links are commonly lumped into a category called automatic surveillance broadcast (ADS-B). These data links could be used to data-link bird targets to the cockpit. The standards, operational procedures and integration with existing avionics systems will need to be upgraded as a long-term solution to the problem of warning pilots of real-time bird activity at and around airports.

The amazing thing about NextGen is that most of the infrastructure has already been developed and is being rolled out. In the United States, most of the East coast and California already have ground-based transceivers installed. General aviation avionics suppliers have already developed the airborne transceivers and displays. So, now I will explain what NextGen is, and at the end of this chapter, I will explain how it can be used to relay the bird targets to the cockpit.

The US aviation industry employs over 11 million people and generates over $1.3 trillion in economic activity each year. Our current radar-based air traffic control system has served us well for the past 60 years but is now outdated, and our busy airports are operated at maximum capacity. Additionally, new aircraft, including very light jets, unmanned aircraft systems, and space transportation vehicles, are being added to an already overtaxed transportation system.

Today, there are more than 600 million passengers, and in 2020 that number is expected to increase to one billion. Without NextGen, there will be gridlock in the skies. By 2020, we estimate that this failure would cost the US economy $22 billion annually in lost economic activity. That number grows to over $40 billion by 2033 if we don't act. Even as early as 2015, a simulation shows that without some of the initial elements of NextGen, delays will be experienced far greater than what we are seeing today. With NextGen, we will move from a ground-based radar system to a satellite-based data link system.

Every airline passenger has encountered delays in their journeys. Some passengers are stuck in airplanes for many hours. The situation is so severe that Congress is currently working on a passenger bill of rights. Our airports are saturated with traffic both on the ground and in the air. NextGen will allow our air transportation to continue to grow while increasing air safety. NextGen will safely allow aircraft to fly closer together on direct routes; thereby reducing delays, and will improve the environment and economy through reductions in carbon emissions, fuel consumption, and noise. NextGen will increase airport capacity, maximize airspace capacity, increase surface efficiency, minimize delays in reduced visibility, allow planes to fly optimal flight paths and improve access. The FAA budget for NextGen has been steadily increasing with a $688 million budget for FY09. The bill for the entire NextGen system approaches $40 billion.

NextGen is the integration of new systems, new procedures and improved aircraft performance. The Joint Planning Development Office (JPDO) was established to support the development of NextGen. The JPDO is made up of representatives from the Departments of Transportation, Defense, Homeland Security, Commerce, the FAA, NASA and the White House Office of Science and Technology Policy. It is also supported by a wide range of aviation experts from across the private sector. The JPDO will develop the concept of operations (CONOPS), which will describe how NextGen will work.

The three primary components of the NextGen evolution are surveillance, communication and navigation. The surveillance technology is called ADS-B as described above and uses GPS signals both to transmit precise aircraft location information (ADS-B Out) for use by air traffic controllers and other aircraft, and to receive precise location information about other aircraft (ADS-B In). This information is provided to flight crews via a cockpit display, which also shows graphical and textual weather information on a moving map.

ADS-B is a satellite-based technology that broadcasts aircraft identification, position and speed with once-per-second updates. ADS-B uses GPS satellite signals to provide air traffic controllers and pilots with very accurate information that will increase safety by improving aircraft separation both in the air and on the ground. Aircraft transponders receive GPS signals and use them to determine the aircraft's precise position in the sky, which is then combined with other data and broadcast out to other aircraft and air traffic control facilities. The system converts that position into a unique digital code and combines it with other data from the aircraft's flight monitoring system, such as the type of aircraft, speed, flight number, and whether the aircraft it is turning, climbing, or descending.

Aircraft equipped to receive the data and ADS-B ground stations up to 200 miles away receive broadcasts. ADS-B ground stations add radar-based targets for non-ADS-B-equipped aircraft to the mix and send all of the information back up to equipped aircraft. This function is called Traffic Information Service-Broadcast (TIS-B). ADS-B ground stations also send out graphical information from the National Weather Service and flight information, such as temporary flight restrictions. This is called Flight Information Service-Broadcast (FIS-B).

As the number of aircraft flying in the National Air Space (NAS) continues to grow and new types of aircraft are introduced, it will be critically important for operators and controllers not only to know precisely where an aircraft is at any given moment, but also where it's going, how fast it's moving, and how long it's going to take reach its destination. NextGen satellite technology will dynamically make this information available to both pilots and controllers with levels of accuracy and precision unattainable by radar. Even though planes will be flying more closely together, the precise information provided by NextGen will significantly increase safety by allowing pilots to know exactly where their aircraft is located in relation to other aircraft throughout all phases of flight.

One of the advantages of the ADS-B system over current radar-based systems is that both pilots and controllers will be able to see the same real-time displays of air traffic. Pilots will have much better situational awareness because they will know where their own aircraft are with greater accuracy, and their displays will show them all the aircraft in the air around them. Pilots will be able to maintain safe separation from other aircraft with fewer instructions from ground-based controllers. At night and in poor visual conditions, pilots will also be able to see where they are in relation to the ground by using onboard avionics and terrain maps. When properly equipped with ADS-B, both pilots and controllers will, for the first time, see the same real-time displays of air traffic, substantially improving safety. ADS-B is currently being implemented nationwide and will provide coverage over the entire US by 2014. The goal is to transition the air traffic control system from radar-based technology to satellite-based solutions by 2020.

A ground-based benefit of ADS-B will be a reduction in the number of ground collisions between aircraft and vehicles. The ground-based system will also reduce what are called 'runway incursions', which occur when an aircraft inadvertently taxis onto a runway without clearance. This situation can occur at airports that are unfamiliar to the pilot. Another incursion scenario occurs when the weather is bad and the pilot has a hard time seeing the runway markings. The new ground-based system is called Airport Surface Detection Equipment, Model X (ASDE-X). The data inputs to this system include ADS-B targets as well as ground-based radar targets. A software tool called Surface Management uses ASDE-X to extend airport surface monitoring beyond runways and taxiways to the ramp areas. This extended coverage will improve common situational awareness between the airport control tower, the airline ramp management towers and the airline operations centers.

NextGen presents the most elegant solution to link bird targets in real time to the cockpit. Once the bird radar detects and tracks the birds in three dimensions, an interface could be developed between the radar and the ground-based ADS-B data link transmitter (1090ES, UAT or VDL-4). This interface could convert the bird radar target parameters (i.e., range, azimuth and elevation) to a coordinate system used by the airliners GPS (i.e., latitude, longitude and altitude). The notice of proposed rulemaking for ADS-B calls for either system to broadcast more than a dozen data elements, including aircraft position, velocity, barometric altitude, and tail number and emitter category. The interface could place the data in the proper format, and the bird target could be given a specific emitter code to differentiate it from other airborne targets. The position and altitude of the bird targets could be transmitted via a data link to the aircraft. The aircraft has a transceiver that will demodulate the transmitted signal and present the bird targets on an ADS-B display that will also display aircraft targets. Aircraft ADS-B displays will need to be modified to allow for the identification and display of bird targets.

In addition to making changes to the ADS-B display, collision avoidance warnings for birds could also be added to the system. Cockpit collision avoidance systems are covered in Chapter 10. Collision avoidance logic could be developed to warn the pilot of an impending collision with a bird; a warning light and audible bird warning could be added to the collection of aircraft collision avoidance warnings. The aircraft present position, programmed route of flight and the ADS-B aircraft targets are shown on a display called the Cockpit Display of Traffic Information (CDTI). The aircraft flight path, bird targets and other aircraft targets could all be integrated onto this single display. A pilot could then see, in a glance, where the birds are in real time. The pilot could also watch the bird target flight path in real time and determine whether it will conflict with the current aircraft programmed flight path. If a conflict exists, a climb, descent or turn could be flown to avoid hitting the birds. So, with this collision avoidance system for birds, a pilot could make real-time decisions to prevent an aircraft bird strike. Currently, very few airlines are using this display, but by 2014, it will be mandatory to have at least the capability to broadcast an aircraft position when flying into controlled airspace.

The first half of this chapter examined airborne operations, but ground operations need to be considered as well. What can be done on the ground to avoid hitting birds on takeoff? I assume that one of two scenarios could unfold. The first would be as the aircraft is taxiing, and just before takeoff, a pilot could monitor the bird targets on the cockpit ADS-B display. When the tower clears the pilot for takeoff and there is a potential conflict, the pilot can simply request a delay until the bird traffic clears.

As mentioned above, one of the advantages of ADS-B is that both pilots and controllers see the same display. If the takeoff occurs at a very busy airport, a delay could have severe consequences. At DFW, where there is one takeoff or landing every 15 seconds, a delay would cause a landing aircraft to go around. So what can be done operationally to resolve a potential bird collision on takeoff? The answer is that the controller will see the conflict as well and can instruct the pilot to fly a different heading on takeoff that will allow him to clear the bird target flight path. Once the aircraft clears the birds, a second heading can be issued to take the aircraft back to its original planned course.

Another system that could be used on takeoff in addition to ADS-B is the ground-based ASDE-X system described above. Although this system is ground-based, it could be modified to add a bird target display mode. There could be a switch for the pilot to either display or not display the bird targets on the ground. After landing, and while taxiing into the gate, the switch would be off. During most of the taxi out, the switch would also be off. The switch would only be on when the aircraft is close to takeoff.

Other Advanced Concepts
Internet Using Wi-Fi in the Sky

Currently, Air Canada, AirTran, Alaska, American, Delta, Frontier, JetBlue, Southwest, United and Virgin offer Wi-Fi Internet access on their airlines. These airlines use one of three Internet services – Gogo (owned by Aircell, the former provider of in-flight telephone service), LiveTV (which JetBlue owns) and Row 44. These services utilize either ground-to-air or satellite data links for Internet access. Gogo offers a ground-to-air data link which operates through a 3G mobile broadband network of 92 ground stations all over the United States.

Back in the 1980s, a company called Airphone started in-flight phone service. In 2000, Verizon acquired Airphone and announced plans to exit this business. The frequency allocation was unused, and eventually it was auctioned by the Federal Communications Commission (FCC), and Aircell won exclusive rights to the frequencies in 2006. In 2008, Aircell used the frequencies to start Gogo as their in-flight Internet portal. The Gogo Internet service turns a commercial airplane into a Wi-Fi hotspot. The current data transfer rates are 3.1 Mbps uplink and 1.8 Mbps downlink. By 2011, the company advertises integration of 4G technology with uplink transfer rates exceeding 300 Mbps.

The Gogo air-to-ground data link uses a widely deployed wireless data technology called Evolution Data Optimized or EVDO Rev A, which is very safe for in-flight use and FAA-approved. EVDO is a wireless standard that uses code division multiple access (CDMA) and time division multiple access (TDMA) protocols, which have been adopted by many cell phone providers around the world. In flight, a cabin telecom router provides a hotspot for wireless devices using the 802.11 b/g protocols. There are several limitations though. First, this service only works over the United States, so if you are an overseas traveler, you will lose coverage once you are over the ocean. The company has announced plans to expand into Canada and Mexico in the future. Blockage by geography and saturation of the ground terminals are other concerns. At approximately $100,000 per aircraft, the integration of the Gogo air-to-ground data link onto an airliner is relatively inexpensive. For passengers, the access costs are currently between $6 for short flights and $13 for a 24-hour pass

Both LiveTV and Row 44 offer satellite-based broadband connectivity. The invention of Ku broadband antennas for aircraft launched the airlines into the Internet business. A DirecTV receiver and entertainment system can be installed on an airliner so passengers can watch TV and surf the Net while airborne. The local TV station links to a satellite ground station by cable or wireless. The ground station has a powerful transmitter, and the TV signal is broadcast to the satellite using an uplink. Large dish antennas are used to provide increased signal strength to the satellite as the signal has to travel 22,800 miles to geosynchronous orbit. The dish is pointed at the satellite, and a signal is transmitted on a particular frequency. The satellite receives and retransmits the signal on a different frequency back to an antenna at a home on earth or to an airliner in the sky. The down-linked satellite signal is normally very weak and requires a series of electronic devices that can amplify the signal to a useable level. Once the amplification is complete, a satellite receiver demodulates the signal and converts it to a usable format.

Row 44 offers Internet, TV and cell phone service (where

permitted). One of the company founders established a relationship with Hughes Network Systems (HNS) to provide in-flight data services to airliners. Row 44 secured an exclusive 15-year contract for North American rights to use the Hughes's Corporation satellite infrastructure around the world. The company leases capacity from the HughesNet satellite Internet access system enabling them to provide worldwide services. Row 44 received a permanent operating license from the FCC. The system consists of a Ku-band antenna mounted on top of the fuselage and a transceiver and wireless LAN installed inside the aircraft. A satellite data link is worldwide and has no geographic restrictions. Satellite-based integration into an airliner is more expensive than ground-to-air at a cost of between $100,000 to $250,000 per aircraft. Obviously, they don't let pilots surf the Net in the cockpit, but using a laptop and wireless access would allow a pilot to see bird targets in the cockpit if the radar published the bird targets to an Internet server. New display systems could easily be integrated into the cockpit that would accept internet inputs.

Internet availability through the data links described in this section could be used as a near-term solution to warn pilots of bird traffic at and around airports. A satellite or ground-to-air data link could be used in an aircraft bird strike warning system in the far term. Initially, pilots could use a separate display for bird traffic. The final solution will integrate the bird radar target into the collision avoidance system and use the existing displays. So, to put the radar and data link together, an interface will need to be designed between the radar and a ground-based computer that performs three-dimensional coordinate conversions. The computer could also host an Internet server. The bird targets could be displayed in three-dimensions on a webpage with an airport diagram overlay. One of the commercial bird radar companies has already integrated bird radar targets onto a near-real-time ground-based two-dimensional Google Earth Internet server at Seattle International Airport as part of the FAA experimental bird radar evaluation program. The bird targets are streamed onto a webpage, and anyone anywhere can pull up the webpage to observe the bird traffic. Since Internet access is already provided to the passengers, wireless Internet access is also available to the cockpit. The pilot could simply use a laptop Internet browser to dial in the Internet address of the bird target display server to observe the bird targets at and around an airport.

In-Flight Pico Cell Phone Technology

While cell phone use in flight is banned in the United States, the European Union approved the use of cell phones on flights in Europe in 2008. Air France, British Air and Ryanair currently allow cell phone use in-flight; however, there is still a ban for use below 10000 feet. There is actually a light in the cabin (like the 'no smoking' lights from the past) that alerts passengers when they are allowed to turn their cell phones on. The service provider for European cell phone use is OnAir. The company uses special equipment to route calls and messages through a satellite network that connects it to a ground-based network. The airplane crew controls the system and can limit or disable its use. Pico cell technology allows passengers to use their own cell phones on planes. Pico cells only serve very small areas though. In ground-based networks, they're used to serve areas such as a building's interior where signals are weak or to boost network capacity. A Pico cell on a plane combines electronic equipment with an onboard antenna. The passengers' calls are sent to the central antenna and relayed to ground towers. In the United States, besides the laws and bans on use, a technical reason for not allowing cell phone use in flight is that the phone would communicate with many ground-based towers on the ground simultaneously. The cellular system was not designed to handle this excess demand, and if many airborne cell phones were used at the same time, the system would most likely saturate. Several airlines are also experimenting with using a cell phone as a boarding pass. The International Air Transport Associations approved a method allowing airlines to send a barcode image to a passenger's cell phone in late 2007. The airline gate attendant can scan the image on the phone the same as a paper boarding pass. This service is projected to start in 2010.

In the United States, we all know that pagers, electronic games, PDAs, MP3 players and CD players must be turned off once the cabin door is closed, until the plane gets above 10000 feet; and cell phones cannot be used in flight at any time while the plane is in the air. But hardly anyone knows the real reason why we have to follow these rules. The FAA fact sheet on cell phone use in flight mentions concerns over many unknowns about the radio signals given off by portable electronic devices (PEDs) and cell phones. These signals, especially in large quantities and emitted over a long time, may unintentionally affect aircraft communications, navigation, flight control and electronic equipment. Therefore, FAA regulations prohibit use of most portable electronic devices aboard aircraft. The rules specifically exempt portable voice recorders, hearing aids, heart pacemakers and electric shavers because they don't give off signals that might interfere with aircraft systems. The regulations also let airlines independently determine whether passengers can use PEDs not specifically mentioned by the rules. An airline must show that a device does not interfere with safe operation of the aircraft during all phases of flight. In its oversight capacity, the FAA ensures that the operator complies with regulations by reviewing the results of the carrier's tests and analysis of pertinent data. The FAA has issued guidance to the airlines allowing passengers to turn on most PEDs after the plane reaches an altitude of 10000 feet. The FAA chose 10000 feet because any potential interference could be more of a safety hazard at a lower height, when the cockpit crew is focusing on critical arrival and departure duties.

Cell phones (and other intentional transmitters) differ from most PEDs in that they send out signals strong enough to be received at distances far from the user. Since 1991, the FCC has banned the in-flight use of 800 MHz cell phones because of potential interference with ground networks. This ban requires that, in addition to testing, the FAA requires an airline to show non-interference to the airplane systems. Before an airline can allow cell phone use in-flight, it needs to apply for an exemption to the FCC rule. Today, airlines may let passengers use newer-model cell phones in what's called "airplane" mode, which essentially disables the transmission function so they can't make calls. This mode lets users do other things, such as play a game, check an address or look at the phone's calendar. In February 2005, the FCC issued a proposal to lift the restriction on 800 MHz cell phone calls while airborne if such phones are operating under control of equipment installed in the airplane that acts as an antenna for onboard callers and controls the power level of the phones themselves. Even if the FCC rescinds its ban, FAA regulations would still apply. Any installed equipment would be subject to FAA certification, just like any other piece of hardware. The air carrier would have to show that the use of a particular model phone would not interfere with the navigation and communications systems of the particular type of aircraft on which it will be used. In addition to FAA and FCC bans, the Halting Airplane Noise to Give Us Peace Act (HANG UP Act) was passed by Congress in 2008 and prohibits cell phone use in flight in the United States. The text of the bill says that *an individual may not engage in voice communications using a mobile communications device in an aircraft during a flight in scheduled passenger interstate air transportation or scheduled passenger intrastate air transportation.*

Since 2003, the RTCA (a technical and operational rulemaking committee) has been looking at the issue of electromagnetic interference from intentionally transmitting PEDs, such as cell phones and Wi-Fi transmitters in laptops. Air carriers have collected data to support the RTCA's work. The FAA is a member of the RTCA committee studying the subject, and reviews data as part of that effort. So far, the data suggests that emissions are well below FCC limits, and are also within limits for interference with aircraft systems. Airlines have been testing and demonstrating transmitting PEDs on a case-by-case basis with the FAA's knowledge. In July 2004, American Airlines and Qualcomm did a successful one-time cell phone test using Pico cell equipment aboard an MD-80 aircraft. In June 2005, the FAA approved United Airlines' request to install equipment for Wi-Fi wireless Internet connections aboard a Boeing 757 aircraft. The approval includes testing to show that the equipment performs its intended function and also that it doesn't interfere with any aircraft systems during all phases of flight.

Internet availability through the data link described in this Section could be used as a near-term solution to warn pilots of bird traffic at and around airports. A cell phone data link could be used in an aircraft bird strike warning system overseas just as in the Wi-Fi adaptation above. As described in the previous Section, the bird targets could be displayed in three dimensions on a webpage with an airport diagram.

The pilot could use a computer with a Pico cell adapter to make an Internet connection. The pilot could then simply dial in the Internet address of the bird target display server to observe the bird targets at and around an airport.

Chapter 14:

Aircraft Collision Avoidance Systems for Bird Warnings

The last existing technology that needs to be integrated is a cockpit collision avoidance system for birds. The technology exists today, but we must treat a bird like an aircraft and then use an aircraft collision avoidance system to avoid a bird strike. Before I cover this topic, though, I want to narrate a bit of the history of aircraft collision avoidance systems. During the Cold War, as a young fighter pilot stationed in Germany, I remember the accidents attributed to collisions that killed my fellow pilots. Fighters do not have collision avoidance systems installed. You must use the mark one eyeball. Being a fighter pilot is one of the most dangerous jobs in the world. The most common accidents I observed were collisions with other aircraft and collisions with the ground. Aircraft collisions usually occurred during formation flying when fighter aircraft are only three feet apart. Formation flying requires intense concentration. Your stomach is tight and you can hear yourself breathing into your oxygen mask. In addition to flying very close to another aircraft, you have to do all the other normal tasks like talking on the radio, checking your fuel and thinking ahead about the mission that needs to be accomplished. Fighters normally assume this close position when the weather is bad to keep the formation together. If you do not fly formation, you may lose track of each other when you enter the weather, which could become very dangerous.

So, what happens when a fighter pilot becomes distracted while flying at such a close distance to another aircraft? How can a distraction happen? What if a light comes on in the cockpit due an aircraft malfunction? What if you lose sight of the other aircraft due to the weather? This actually happened to me as a new wingman in Europe. I took off on the wing of my flight lead. We went into the weather, and poof—my flight lead was gone! The clouds were so thick it was impossible to see. In training, you are taught to execute a procedure called 'lost wingman' at times like these. You simply turn away from the leader's last known position, hold that heading for a set period of time, and when you gain enough separation, either call the controller and ask for a heading to rejoin, or use your onboard radar to find your flight lead. As you can imagine, the whole experience is very disorienting. Not only are you trying not to hit another aircraft, you are basically doing this blind. There is never any chance to practice either; you are academically trained to produce an automatic response. So, when it actually happens, there is initially some confusion. When this maneuver is not carried out with enough precision, collisions can and do occur. I have seen and read about many pilots losing their aircraft and their lives due to a mid-air collisions in scenarios such as this.

Fighter aircraft can collide in other ways, too. When practicing air-to-air combat, a fighter engagement scenario starts with 20 to 30 NM of separation. The aircraft then turn into each other and call "fight's on." The game is to find the opposing aircraft on the radar, employ an air-to-air missile (simulated in peacetime, of course) and go after the next guy or go home. Well, while you're proceeding toward each other with closure rates in excess of 1,000 miles per hour, you are required to maintain altitude separation. The whole engagement happens in less than about two minutes. A pilot maintains altitude separation by referencing his altimeter and making sure he does not drift into the other aircraft's altitude block until he sees him. Well, guess what happens – about a hundred things are going on in a very short period of time, and a pilot sometimes forgets to reference his altimeter. As a result, he drifts into the other aircraft's altitude block and whamo – he hits the other aircraft and pieces of airplane rain onto the ground.

In addition to air-to-air collisions, aircraft sometimes collide with the ground. In the military, you're trained to fly at very low altitudes so that you can fly under a radar's coverage to avoid detection. When an aircraft is very close to the ground, though, there isn't much time for a pilot to react when distracted, and sometimes the aircraft hits the ground (at 100 feet you have 6 seconds). As the saying goes, the ground has a PK (probability of kill) of 100% (It's actually 99% because one guy actually survived a collision with the ground by hitting the ground at a very low graze angle and skipped off). Another way an aircraft hits the ground is during instrument approaches in the weather during the landing phase. A pilot watches his instruments as he descends in order to land while in the weather. For every airport, there is a fixed altitude that the aircraft cannot go below unless the runway is in sight (airlines have other requirements because they have more precise landing systems). There is a very critical phase of flight when the pilot must make a decision to either land or go around. If the aircraft breaks out of the weather and the runway is not in sight, the pilot will normally make a decision to push up the power, clean up the aircraft and go around; but if the wrong decision is made and the pilot continues without having the runway in sight, he may hit the ground and have a very bad day.

During my military flying years, I saw many of the above scenarios happen. When I was stationed in Germany, I soon discovered that part of the planet has some of the worst weather in the world. Young fighter pilots are normally trained at bases around the country where the weather is very good for most of the year. Immediately after training, a fighter pilot is often sent overseas to his unit to receive advanced training to qualify as a combat-ready fighter pilot. Guess what – the results have sometimes been disastrous. Young fighter pilots have become disoriented in the weather and bailed out and crashed into the ground because of misjudgments during approaches to land.

Fast forward to my time as an airline pilot. It turns out that airliners had these same problems of hitting other aircraft and hitting the ground. Several very serious accidents led Congress to pass a law requiring airliners to have collision avoidance systems installed. The first was a collision in the Grand Canyon in 1956. Two airliners collided, resulting in a crash of both planes and 128 fatalities. At that time, it was the deadliest aviation disaster in aviation history. Investigation of the accident was difficult because CVRs and FDRs were not in use at that time. Also, at that time, the NTSB did not exist and the Civil Aeronautics Board (CAB) reached the following conclusion: "The Board determines that the probable cause of this mid-air collision was that the pilots did not see each other in time to avoid the collision. It is not possible to determine why the pilots did not see each other, but the evidence suggests that it resulted from any one or a combination of the following factors: intervening clouds reducing time for visual separation, visual limitations due to cockpit visibility, and preoccupation with normal cockpit duties, preoccupation with matters unrelated to cockpit duties, such as attempting to provide the passengers with a more scenic view of the Grand Canyon area, physiological limits to human vision reducing the time opportunity to see and avoid the other aircraft, or insufficiency of en-route air traffic advisory information due to inadequacy of facilities and lack of personnel in air traffic control." When the public discovered that the technology used for air traffic control was outdated, they became outraged over this accident. Congressional hearings occurred (as with Flight 1549) in 1957, and the result was the Federal Aviation Act of 1958, which created the FAA, upgraded technology, added more controllers and gave the FAA authority over all US airspace. Although the number of mid-air collisions decreased statistically since then, there was still no permanent technical solution to the problem.

Another serious airborne collision between two aircraft occurred in 1978. An airliner and a small private aircraft collided over San Diego, resulting in what was the deadliest airplane crash in US history at the time with a death toll of 144. The NTSB investigation revealed that the airliner had been instructed to maintain visual contact with the small aircraft but lost sight and failed to communicate this to ATC. A result of this accident was the creation of Class B airspace. In 1980, the FAA created this airspace, which prohibited visual separation of aircraft and instead required positive radar control when operating an aircraft close to a very busy airport. The final accident that prompted Congress to pass a law requiring the installation of collision avoidance systems occurred over Cerritos, CA in 1986. Again, an airliner and a small private aircraft collided in flight, this time resulting in 82 fatalities. The NTSB investigation discovered that the small aircraft was not in communication with the controller and had no transponder (which indicates an aircraft's altitude, among other things).

The concept and initial research and development for aircraft collision avoidance systems dates back to the 1950s. Development efforts were continued into the 60s, but a producible system was never realized, primarily due to the unreliability of data links. Later in the 1960s and early 70s, development efforts led to producible systems; however, they were never adopted due to their high false alarm rates. After leaving the military, I spent some time at MIT Lincoln Laboratory as a radar engineer. This lab is a Federally Funded Research and Development Center (FFRDC) and receives about $200 million per year from the FAA for research and development. While I was there, I shared an office with a very smart hardware guy, Bill Petrovick, who had done some work on an aircraft collision avoidance system in the 70s called Beacon Collision Avoidance System (BCAS). BCAS used aircraft transponders and ground interrogations to calculate nearby aircraft locations and was used primarily in the general aviation market.

In 1981, the FAA started a program called Traffic Collision Avoidance System (TCAS). The FAA took on this project and was responsible for conducting research, development, testing and demonstration, and defining the technical and operational feasibility of the concept (I believe now is the right time for the FAA to take on an aircraft bird collision avoidance project). There were many difficulties to overcome in the TCAS technology development program, including antenna design and placement, clutter, false alarms and operational implementation. The resulting antenna design was one that sends out a signal in all directions (360 degrees) simultaneously. Software advances eventually overcame the problems of clutter and false alarms (In the same fashion, we need to deploy a prototype aircraft bird collision avoidance system so we can solve the technical and operational issues).

In 1986, after five years of R&D on TCAS, a small private aircraft entered into Los Angeles airspace without talking to the controller. Despite the fact this it was a clear day, the small aircraft collided with an airliner, killing all onboard and many on the ground. The NTSB investigation discovered that the small aircraft had entered controlled airspace without communicating with the controller and with no transponder installed. Air traffic control could see the aircraft on their screen, but because the radar was only two-dimensional, the controller could not see the aircraft's altitude. What's more, there were no automatic warning systems in the control facility (IFF was added at a later point to add altitude as the third dimension).

Before 1986, as a result of this and other aircraft in-flight collisions, the NTSB published a total of 14 recommendations for installation of a collision avoidance system on airliners. As a result of this accident, Congress passed the 1987 Airport and Airways Capacity Expansion and Improvement Act, Public Law 100-223 was passed, establishing deadlines for completing development and installing collision avoidance systems on all commercial airliners. All commercial aircraft with more than 30 seats or a maximum takeoff weight greater than 33,000 pounds were required to be equipped with collision avoidance systems.

It is interesting to note that it was not until 1990 that the first commercial collision avoidance system was installed, and it took until 1993 for all airliners to complete the installation. There are many similarities between the events leading up to the development of collision avoidance systems for airliners and the significant accidents that have occurred over the last 40 years involving aircraft collisions with birds. And although Appendix 4 lists 26 NTSB recommendations on bird strike avoidance, none of these concern a collision avoidance system for birds because such a system has not yet been invented.

TCAS uses interrogations and replies from the aircraft transponders and a collision avoidance system to display air traffic in the vicinity of the aircraft. The interrogation rate is once per 5 seconds when the threat aircraft is 5 NM away, and once per second when the aircraft is very close. An airliner TCAS can interrogate a small aircraft transponder, and the cockpit display shows the aircraft position and altitude. With TCAS, the pilot has advance warning on another aircrafts' three-dimensional position and can maneuver in advance to avoid a collision. This is very similar to the way a bird collision avoidance system could function. A pilot could see bird tracks in advance and maneuver to avoid them. It is also interesting to note that the TCAS system is independent of air traffic control. A pilot does not need a controller to tell him where nearby aircraft are because they would already be displayed in the cockpit. As a matter of fact, the transponder is sometimes called a 'secondary surveillance radar' (SSR). For a complete description of TCAS installation and use, see Appendix 7.

In addition to aircraft carrying transponders and interrogating each other, there are ground-based interrogators in the control tower. Airborne aircraft are interrogated by the ground-based interrogator, and the aircraft target's transponder returns are displayed on the controller's display. The TCAS system actually sends out an interrogation signal several times per second to all aircraft on one frequency and receives a reply from all aircraft equipped with a transponder on another frequency. The TCAS then saves all of the aircraft range, altitude and bearing data in memory. The TCAS algorithms then predict future aircraft locations and determine whether the potential for a collision exists. If a collision potential does exist, then the TCAS system determines an avoidance maneuver for the pilot to fly. Currently, aircraft are only allowed to climb and descend to avoid a collision because of lack of azimuth accuracy. No turns are allowed unless there is visual confirmation with the other aircraft. The pilot receives both a visual and audible warning to avoid a collision.

The TCAS system has controls that allow the pilot to select multiple ranges and altitude bands for traffic to be displayed. Longer ranges are normally selected in cruise flight when there is greater separation, and shorter ranges are selected in the terminal areas near an airport due to higher traffic densities. The symbology and audible and visual warning systems are used to warn the pilot of an impending collision. Non-threatening traffic within 40 NM are displayed with an open white diamond on the navigation/radar and/ or vertical velocity displays (each aircraft has made their own efforts to integrate the TCAS system into their existing displays). When the traffic is within 6 NM and +/- 1200 ft, the hollow white diamond is filled in. When the traffic becomes close or has parameters that exceed certain thresholds, then the white diamond becomes a yellow diamond and an audible warning says "traffic, traffic". This is called a 'traffic advisory' or TA. The symbol is displayed relative to the aircraft. At this time, the situation is still advisory, mild, and simply requires a monitoring of the potential conflict. The pilot's response is to look at the TCAS display and determine where the conflict is relative to the aircraft. The potential conflict's range, bearing and relative altitude are displayed along with a climb or descend indication. If the weather is good, then the pilot can look around to gain visual contact with the conflict.

When the parameters for a potential conflict become critical, the yellow diamond becomes a red square and an audible warning commands the pilot to perform an avoidance maneuver. This is called a 'resolution advisory' or RA. This command will be either "climb, climb" or "descend, descend". If the other aircraft is equipped with TCAS, the opposite command will be given for the avoidance maneuver. In other words, one aircraft will be given the command to climb and the other aircraft will be given the command to descend. So, if you follow the calculated and commanded actions of the TCAS, a potential collision will be averted. The TCAS maneuver is normally hand-flown by the pilot, which means that if the autopilot and autothrottles are engaged, the pilot must first disengage these systems. After disengagement, the pilot flies the maneuver, and when the TCAS command "clear of conflict" is broadcast, the pilot returns the aircraft to its previous flight conditions. Other commands given by the TCAS can be to "monitor vertical speed" or "maintain vertical speed", which means not to climb or descend. Other information displayed on the TCAS display includes the threat aircraft altitude in hundreds of feet and a + or – sign. The + means above and the – means below. Additionally, an up arrow means that the threat aircraft is climbing and a down arrow means the threat aircraft is descending. It is also interesting to note that, normally, all maneuvering in an aircraft (i.e., turns, climbs and descents) occurs only with the controller's permission. There is a very strict discipline between pilots and controllers. When a controller tracks a pilot on his radar, he directs the pilot to perform a certain maneuver, depending on the phase of flight (i.e., takeoff, cruise, descent, approach and landing). Normally, the only way for a pilot to deviate from a controller's commands is to declare an emergency. When a TCAS RA occurs, pilots are authorized to deviate from the controllers clearance to prevent a collision.

Many years are normally spent vetting a new system. In the early 1980s, TCAS was a new technology, but the question at the time was how does one employ the system operationally (very similar to the current FAA bird radar evaluation program). It took 10 years to create the first operational TCAS. When a new technology is introduced into the National Airspace System, various rulemaking committees convene to determine the optimal operational implementation. These committees include FAA, ALPA, RTCA, NASA, commercial avionics companies, airlines, mechanics, etc. Their purpose is to determine how to integrate, operate, maintain and train personnel for the new technology. In addition to operating the new system, maintenance inspections are needed to ensure safe operation. Inspections are normally performed on an hourly basis. For example, maybe every 200 hours, a ground-based test will need to be performed on TCAS to ensure operation. So, one of the operational procedures decided by the committee for TCAS was to give the TCAS warning an emergency priority. In near-collision scenarios, pilots are authorized to deviate from a previous clearance to follow TCAS commands and indications. Currently, TCAS upgrade is an ongoing area of research for improvement in air safety. There is, however, a next generation of TCAS (TCAS III) that will allow horizontal maneuvers (i.e., turns) to occur. This type of maneuvering will increase the separation and therefore the safety margin between aircraft. These improvements are currently research only, though, with no plans for implementation.

It is interesting to note that when a new technology is integrated and used for a while, the operators sometimes come up with useful applications. When I have flown into very busy airports like JFK, there are normally numerous aircraft ahead of me and behind me that I need to be conscious of so that I don't screw up the controller's sequencing. The controller plans a separation between aircraft so that there is enough time to land and clear the runway before the next aircraft lands. As an aircraft is approaching to land, it slows down and configures (i.e., lowers the gear and flaps). As one aircraft slows down, the aircraft behind it catches up. When an aircraft lands, the runway must be clear of traffic. If the aircraft behind is too close and the aircraft in front is still on the runway, then a go-around must be accomplished. A go-around usually means the aircraft must climb out and go all the way around the traffic pattern and accomplish another approach and landing.

Another useful application for TCAS is during ocean crossings. There is no radar coverage over the ocean, so a pilot must report his position on the radio to a controller on either side of the ocean. The controller keeps track of all reported aircraft positions and attempts to control traffic this way. Modern commercial aircraft actually have SATCOM data links that automatically report their positions, but even with this automation, there is still a requirement to maintain high-frequency radio contact with a controller. Over the ocean, TCAS (in conjunction with on board weather radar) has been used to navigate around thunderstorms. You can see the multiple aircraft in front of you changing course to avoid the areas of intense precipitation. You are in effect using the decisions of other pilots to create a navigation solution. In the same way, an aircraft bird strike collision avoidance system prototype needs to be developed so that pilots can create an operational use that will allow them to avoid hitting birds.

Another serious collision avoidance problem for the airlines was controlled flight into terrain (CFIT). This occurs when an aircraft hits the ground during low-altitude flight. The collision normally occurs unexpected by the flight crew (loss of situational awareness, which means the pilot is not exactly sure where he is) and mostly during the approach and landing phases of flight. In the late 1960s, there were numerous accidents where a perfectly good aircraft hit the ground and hundreds of people were killed. The accident that finally spurred change was a TWA flight into Washington Dulles in 1974. The flight crashed into Mount Weather, VA. The NTSB found the probable cause of the accident to be the flight crew's decision to descend to 1,800 feet before the aircraft had reached the approach segment where the minimum altitude applied. However, two of the five-member Board dissented, identifying the probable cause to be the failure of the controller to issue altitude restrictions in accordance with the terminal controller's handbook. The pilot was also cited for failure to adhere to the minimum sector altitude depicted on the instrument approach chart and failure to request a clarification of the clearance. So, in this accident, both the pilots and controllers were faulted.

In most cases, either the pilot, the controller or both make errors that contribute to a CFIT accident. Pilots often encounter bad weather with very low clouds in an unfamiliar area. Controllers normally give pilots a heading to fly for various phases of flight. During the approach phase, the controller gives the pilot a heading to intercept an instrument approach that will guide him through the weather to a landing. Sometimes, controllers are distracted because they are controlling more than one flight. If there is terrain and the controller forgets to give the pilot a turn at a critical time, the aircraft could impact terrain. Another scenario that has occurred is impact with the ground during a descent. Normally, an airliner descends from a very high altitude using a series of stepdown altitudes until the aircraft is on the instrument approach. Accidents have occurred where the controller has directed the pilot to descend to an altitude lower than the terrain. Because there is a trust between controllers and pilots, the directions are followed and the aircraft impacts the ground. Pilots are now taught to be aware of the minimum vectoring altitude or the minimum altitude an aircraft should be at for the particular part of the planet that the aircraft is over.

In addition to pilot and controller mistakes, another factor contributing to the TWA accident mentioned above was a lack of crew coordination. Crew Resource Management (described in Chapter 1) had not been invented yet. One of the pilots was uncomfortable with where the aircraft was relative to where it should have been, but at that time, no one questioned the Captain's decisions. Today, though, aircrews are trained to work together, and when there is a question or something does not feel quite right, pilots are taught to speak up. After this accident, the NTSB recommended and the FAA required airlines to install Ground Proximity Warning Systems (GPWS). Since 1974, many improvements have occurred, and the current system is called Enhanced Ground Proximity Warning System (EGPWS). Statistics have shown that, since EGPWS was installed, there have been no accidents that were due to CFIT.

The inputs to EGPWS are the radar altimeter, digital terrain database, GPS, airspeed, altitude, air data computer, and landing gear and flap configurations. These inputs are used by alert algorithms to generate aural and visual warnings. The aircraft's present position and altitude is monitored along with the surrounding terrain. All of these inputs allow the EGPWS computer and its corresponding software to calculate the type of warning given depending on the phase of flight (e.g., cruise at low altitude, approach to land, and takeoff). When the aircraft is cruising close to the ground and there is a potential collision with the terrain in the projected flight path of the aircraft, the audible warning "terrain, terrain, pull up" will occur. Additionally, if terrain monitoring is enabled, a color-coded image of the terrain provides situational awareness. In order to avoid a collision with the terrain, pilots are trained to disconnect the autopilot, advance the power, and pull up as hard as possible without stalling the aircraft.

When an aircraft is descending to land and is close to the ground and the descent rate is excessive, the audible warning "sink rate" is generated. If the descent rate is severe relative to the aircraft altitude, the warning "whoop, whoop, pull up" will be heard. The pilot reaction in both of these cases is to reduce the rate of descent until the warning subsides. If the aircraft is close to landing and a pull-up does not occur, it is possible that the aircraft will crash onto the runway and you will have a very bad day. Another warning occurs when the aircraft is approaching to land and the gear or flaps are not down. There are various conditions that could cause this warning. The warning "too low gear" or "too low flaps" could occur. The pilot reaction, in this case, is to either disregard the warning or go around for another approach. Another condition that could activate the EGPWS is a deviation from the ILS glide slope. ILS is an acronym for 'instrument landing system'. An ILS is a precision approach that guides an aircraft down a precise descent path to a landing. If the aircraft deviates significantly below the glide path, then the audible warning "glide slope" occurs. Another phase of flight for possible collision with terrain is during takeoff. If the aircraft takes off, and, instead of climbing, it descends, then a "don't sink" warning will be heard. The pilot should look at his altimeter and vertical velocity indicator and correct the problem by pulling back and changing the descent to a climb.

Now that you are an expert on TCAS and EGPWS and their operation and implementation, I will attempt to extrapolate how a new collision avoidance system could be used to prevent bird strikes. Since this is a new invention, there are no operational procedures or CONOPS for the procedures that pilots would follow if they had an indicator for bird targets. Various agencies will need to come together to define standards and operational procedures so that the same indications occur on a cockpit display system no matter who the manufacturer is. This standardization will also allow pilots to be trained more easily because the indications will be the same no matter which aircraft is flown. The FAA needs to task RTCA to provide consensus based technical recommendations and operational procedures to resolve complex aviation issues.

RTCA membership includes over 335 government, industry and academic organizations from the United States and around the world. A sample of RTCA membership includes the Federal Aviation Administration, Air Line Pilots Association, Air Transport Association of America, Aircraft Owners and Pilots Association, ARINC Incorporated, Avwrite, The Boeing Company, Department of Defense, GARMIN International, Rockwell International, Stanford University, Lockheed Martin, MIT Lincoln Laboratory, MITRE/CAASD, Harris Corporation, NASA, National Business Aviation Association, and Raytheon. RTCA is supported by many international organizations, including over 100 International Associates, such as Airservices Australia, Airways Corporation of New Zealand, Airbus, the Chinese Aeronautical Radio Electronics

Research Institute (CARERI), EUROCONTROL, NAV Canada, Bombardier Aerospace, Society of Japanese Aerospace Companies, Thales Avionics Limited, Centre for Airborne Systems-Bangalore, the United Kingdom Civil Aviation Authority and many more.

So now that you understand the technical and operational operations of TCAS and EGPWS, it is time to move on to how this technology, along with bird targets in the cockpit, could be used to prevent airline bird strikes. I will now present different operational scenarios and possible cockpit warnings and pilot procedures that could be accomplished to avoid a potential aircraft bird strike. It is assumed that a ground-based sensor has detected and tracked the bird targets. It is further assumed that there is an interface between the sensor and the ground-to-air data or ground-to-space-to-air data link. It is also assumed that standards have been developed.

Since aircraft targets are already displayed in the cockpit, it will be necessary to create a separate code and symbol to differentiate birds from aircraft targets. The bird target three-dimensional position will need to be stored in memory. A software algorithm will then use the stored positions of the aircraft and bird targets to make a projection for a potential collision. The bird targets and their projected flight paths could be overlaid on the navigation display, thereby allowing the pilot to observe the future aircraft navigation path and determine whether a future conflict would exist with a current or projected bird target track.

As for operational implementation, I would divide this problem into three parts according to the phase of flight – takeoff, cruise and landing. To calculate the proper avoidance maneuver, software algorithms will need to be created for each phase of flight. The takeoff phase is probably the most critical phase of flight because the aircraft is slow and difficult to maneuver. I would estimate that there is almost no time to react once the takeoff roll is started. The last time a pilot can really make a decision if he sees the bird targets on a display is when he is taxiing onto the runway. Once the takeoff roll is started, though, there is little value to a display with bird targets on it. So how would a pilot use the display on takeoff? I believe the answer to this question will be determined after operational evaluation. When the aircraft is on the ground and ready for takeoff, if the pilot observes a potential conflict, he will have to project where the bird track will be at some future point in time and determine if a potential conflict exists. This can be done with software in an aircraft bird strike warning system. Departure and arrival routes are displayed in the cockpit, and they are not straight lines off the projected centerline of the runway. The software needs to compare the programmed navigation path of the aircraft to the projected flight path of the birds. If there is a future potential conflict, the pilot could be warned with a visual color-coded symbol and an audible warning. Once the pilot sees a potential conflict, he could simply request a short delay from the controller to allow the birds to fly through or past the area of conflict. After the delay, the pilot could request takeoff clearance.

The takeoff scenario described above will most likely work at low-traffic airports. Airport traffic capacity is measured in operations per hour, and an 'operation' is defined as a takeoff, landing or go-around. At very busy Class B airports (i.e., airports where the annual traffic count exceeds five million passengers), airline capacity can exceed over 240 operations per hour or one operation every 15 seconds. If a pilot asks for a delay for birds at a high-traffic capacity airport, the controller has to worry about the landing traffic. A delay could cause an aircraft on landing approach to go around because it can't land with an aircraft still on the runway. So how does one solve this chicken-or-egg problem?

One possible solution is to simultaneously data-link the bird targets to both the pilot and controller. On takeoff, the controller could see the bird target conflict and change the departure instructions so that the aircraft taking off will miss the birds. For example, let's say the aircraft taking off is making a left turn after takeoff and the controller sees a conflict. The controller can simply say "after takeoff maintain runway heading." The controller will obviously have to de-conflict this change, but this is a simple example of how technology and procedures for pilots and controllers can prevent a bird strike.

So far, I have described prevention procedures for the takeoff phase of flight, but what about the landing phase? Suppose a pilot is approaching to land and sees a potential conflict. What can be done to prevent a bird strike at this point? Well, a pilot has several options.

One is to simply go around. If the pilot knows the altitude of the birds, he can simply push up the power and climb to an altitude above the birds to avoid a collision. The aircraft will then be flown around the traffic pattern to attempt another landing. By the time another landing is attempted, the birds will be well on their way.

Another possibility during the approach phase is to have the pilot ask the controller for a delay until the bird traffic is clear of the flight path. One type of delay is a vector, which is just a heading change. A heading change followed by a second heading that will place the aircraft back on course will be enough of a delay to avoid a collision. The nice thing about this approach is that either the pilot or controller can initiate the heading change depending on their display observation. A second possibility for a delay is to ask for a turn in holding. When this maneuver occurs, the aircraft simply makes a 180-degree turn. This turn will cause enough of a delay to allow the birds to move on and avoid a potential collision. If this maneuver is conducted in high-traffic density airspace, it will cause a ripple effect delaying the aircraft behind; however, the priority of safety prevails, and the aircraft behind will just have to wait.

The maneuvers described thus far represent my personal opinion as a major airline pilot, but I am not naïve enough to think that I have the only solution; clearly, two heads are better than one. I believe it is important to have controllers, pilots, the FAA, the RTCA, NASA and many other organizations create a comprehensive solution to determine the correct procedures for a pilot to accomplish. In fact, this is exactly what RTCA and other organizations do – they bring together the right mix of people to solve the problem. Initially, the technology will be deployed at airports, and as operators and controllers use it, they will encounter problems. These problems will be identified, investigated an analyzed to determine their causes. Solutions will be proposed, simulated, implemented and tested. The government will oversee the entire process, and decisions will be made to finally incorporate the change.

Thus far, we have covered the takeoff and landing phases of flight, so what remains is the cruise phase, which is much less threatening because the aircraft is much further away from the ground and there is more time to react. If a pilot were to see a potential conflict at let's say 4000 feet, it's a very simple matter to ask the controller for a change of heading to prevent from running into the birds. If the bird target heading is also available to the controller, then a different heading can be issued to the pilot to avoid the birds. If the birds are at the same altitude as the aircraft, then the pilot can simply climb or descend to avoid a collision. I can almost guarantee that if another Flight 1549 occurs where both engines are lost due to birds and the aircraft comes down (most likely with severe consequences), we will have another Congressional law passed as was done with TCAS. Currently, though, there is no recommended long-term solution to this problem – there is no technical solution in the works. Aviation operations are very difficult to explain. My solutions make immediate sense to pilots, but to an engineer who has never flown, they are difficult to comprehend. There are those that get it and those that don't. I believe that the optimal technical solution to the aircraft bird strike problem is to use ground-based radar to detect the birds and data-link the targets simultaneously to the cockpit and the tower. Changes must be made to the aircraft collision warning systems to incorporate bird targets as a potential collision hazard. This will allow users to make real-time decisions to prevent bird strikes.

So there you have it. A complete collision aircraft avoidance system for birds composed from technological components that are currently available – a bird radar, a data link to the cockpit, a collision avoidance system, an integrated display and operational procedures. To solve this problem which has caused immeasurable damage and loss of life, the implementation will need to be spurred by government investment and a consortium of FAA, airlines, operators, maintainers and commercial avionics developers. Back in the 80s, we fixed the problem of aircraft hitting other aircraft and hitting the ground. Now let's put an end to this problem once and for all.

Chapter 15:

Recommendations

Judging from the fact that the number of aircraft bird strikes is increasing, it is safe to say that the current approach of catching, shooting, poisoning and scaring birds at and around airports is simply not working. Even if this approach is improved, the number of aircraft and birds fighting for the same volume of airspace is increasing at a dramatic rate. The current goals of the FAA bird radar evaluation project are to determine expected radar performance; validate target quality; understand the impact of the local environment on system performance; assess electromagnetic compatibility with other airport systems; assess data management, integration and interoperability; work with wildlife management personnel for proper implementation; establish minimum requirements; determine updates to regulations; and develop guidance on how to acquire, deploy, integrate, acceptance test, operate and maintain bird radars.

The FAA started its bird radar research in 2000 in an effort to determine whether low-cost radars can reliably detect birds at or near airports (i.e., 3 to possibly 5 miles) and be used to develop an airport bird strike advisory system. The radar data is currently overlaid onto an airport geographic information system (GIS). It is not a stretch to say that radar data could soon be displayed at the airport operation center or possibly in the aircraft cockpit. The FAA knows that the radars can see birds, but they don't know if the data is useable by a controller or pilot to make a bird avoidance decision in real time. Their bird radar may be of most use to airport operators who use it to manage their wildlife control programs. The research is continuing to address these operational-type issues.

Currently, the FAA is conducting radar evaluations with two bird radar systems at Seattle-Tacoma International Airport, two bird radar systems at Naval Air Station Whidbey Island in Oak Harbor, WA, and one portable research radar unit that is owned/leased by the University of Illinois (CEAT) and currently finishing a brief deployment at YVR (Vancouver, BC Canada). The next steps for the FAA are to conduct additional testing at Chicago O'Hare International, Dallas Fort Worth International and John F. Kennedy International Airports.

While I believe that bird radars will be a great tool to help wildlife personnel do a better job controlling birds at airports, this technology falls short of a permanent technical solution. No long-term plan has been initiated to eradicate the bird strike problem once and for all. In order to resolve or reduce the growing threat of aircraft bird strikes, I would like to make the following recommendations:

Allocate Long-Term Funding to Develop a Collision Avoidance System for Birds

The FAA's current strategy shift from R&D to evaluation of commercial bird radars as a project is a good start. Their current goal is to define technical requirements and determine how to acquire, operate and maintain bird radars at airports. Minimum requirements for radar design should be defined as soon as possible so that commercial companies can begin to invest in robust designs. Recommended requirements include 3-dimensional coverage over 360 degrees, including vertical coverage from surface to 3000 feet and horizontal range to 5 NM. The radar should be designed to determine bird target altitude, velocity, heading, size, species (can't be obtained using a radar but integrating an optical sensor with the radar could provide this capability), and projected flight path. A radar design should include a storage unit that writes certain bird target parameters to a database. Birds tend to use the same migration flight paths, and storing the bird data would allow wildlife biologists to work with air traffic managers to move departure and arrival flight paths to avoid birds (possibly on a seasonal basis). The database may even be used as a source of data for NTSB bird strike accident investigations. In the short term, risk levels for bird activity should be quantified and automated so that wildlife personnel can report the current threat level to the tower and so that it can be passed on to pilots through ATIS. Certification standards should be developed and enforced. A final requirement is low cost.

Another area that should be added to the current FAA bird radar evaluation is operational implementation. As members of the evaluation committee, technical Federally Funded Research and Development Centers (FFRDCs) that specialize in radar research and development should be included. The FAA has declared current radar technology as not being operationally suitable for a pilot to make a decision; however, the technology will mature and reach that point in the future. I believe the users should be included now and should be involved in defining the final requirements for operational use of airport bird radars. Bird radar design should be viewed as a first step in developing a sensor system that will data-link targets to the cockpit. The radars, however, won't be capable of preventing bird strikes on their own; pilots will play an integral role in any solution. As a minimum, airport wildlife biologists and specialists as well as air traffic controllers and pilots should be made members of the evaluation committee. User membership would go a long way in soothing recent tensions with ALPA, which has publicly expressed disapproval for the slow development of a comprehensive solution.

Within a year, the FAA will educate airports on bird radars through a variety of means and approve federal funding for their purchase. An affordable bird radar is currently being evaluated as a tool for airport wildlife specialists to see birds in the sky at distances beyond visual range – a capability that did not exist until just recently. While the FAA is working with a limited number of wildlife personnel for implementation, there is no CONOPS for bird radar employment. This means that when airports buy the bird radar, they will have no operational procedures for their employees.

Each airport will be trying new ways to use the bird radar as a tool. In light of this fact, I deeply believe that a group of users (like RTCA) should form a committee to develop the CONOPS ahead of the deployment of bird radars at airports.

The above description is the most likely path that the FAA will take to reduce the risk of an aircraft collision with a bird. I believe that, in parallel with this effort, the FAA should invest in a collision avoidance system for birds as described in this book. In my experience, a technical demonstration can go a long way in convincing those in doubt that the concept actually works. To jumpstart the investment, a simple demonstration could be performed. Simply publish the bird targets from a commercial bird radar to an Internet server, and use an aircraft with wireless installed to display the bird targets on a laptop in the cockpit. Contact the airlines, and create a voluntary evaluation program with pilots flying with bird radar engineers in the jumpseat while the aircraft is in revenue service. Have pilots fill out surveys after the flight, and use the survey as a user evaluation tool. Bird targets in the cockpit, at this point, would be advisory only, and the pilot would not need to maneuver the aircraft in relation to the birds. A simple survey report could be written after each flight to explain overall effectiveness. During this same demonstration, the control tower could use a computer client to display the same bird targets. Use of the Internet to display bird targets will allow the pilot and controller to see the same targets at the same time. Controllers could provide pilots with an advisory warning for the location of birds on a time-available basis. In parallel with this demonstration and evaluation, the government (i.e., FAA, USDA, USACOE, USEPA, USAF) should invest in a complete bird strike collision avoidance system – the bird radar, the data link, the displays, the collision avoidance algorithms, the avionics and integration into existing or new cockpit displays. To provide a permanent technical solution once and for all, this should be an ongoing long-term investment.

Develop CONOPS for Bird Radars to Warn Pilots & Controllers

Currently, CONOPS for bird strike warnings start with wildlife personnel inspecting the runways and airport environment for bird activity. During the inspection, a visual determination of the threat level (i.e., light, medium, heavy) is made. When the threat level changes, a call is made to the control tower. The control tower then updates the Automatic Terminal Information Service (ATIS), which pilots can consult before takeoff and before landing. As mentioned before, this report is not very useful to pilots, though, because it does not specify where the birds are. The current thinking for implementation of bird radars as tools for wildlife personnel are to help in the assessment of the risk level. Instead of using the eye at short ranges, the thinking is that a bird radar will be able to see birds at greater ranges and automate the function of threat level determination. Once the radar reports the threat level, the same unusable process of making the call and updating ATIS is performed.

As I learned through my experience with military acquisition, CONOPS and missions directly affect the requirements for the design of weapon systems. When, for example, the government creates a mission for an aircraft to kill tanks, it first decides on using a 30mm Gatling gun that shoots 6,000 rounds per minute (or 100 rounds per second). These requirements directly influenced the design of the A-10 Warthog. When prisoners from the first Gulf War were interviewed, they said that the A-10 was "the most feared aircraft". In the same manner, as government acquisition programs are conducted, bird radar requirements should be designed for a mission or CONOPS. There should be a long-term plan for bird radar development as a sensor in a bird strike collision avoidance system.

CONOPS development for use of an airport bird radar will need to take a phased approach depending on where the bird targets are displayed. Phase one would consist of training the airport wildlife specialists to operate and understand the bird radar display systems. This can be done as soon as an airport purchases a bird radar. A wildlife specialist could observe a potential bird strike conflict and advise the controller, who can then, in turn, advise the pilot of the location of the birds. When wildlife specialists are called out to scatter birds near a runway, they could watch the dispersal on the radar and track the targets to make sure that no conflicts with aircraft occur. Once the targets are clear of the runway environment, the wildlife specialist could notify the control tower that the coast is clear for takeoff or landing.

No North American Airport control tower current has a bird radar display in it. In fact, a bird radar display is not presently allowed in any US FAA control towers. As of the writing of this book, only the military, NASA and one airport in Africa have allowed a bird radar display in the control tower. The CONOPS for their use is unknown, but I do know that the number of operations (i.e., takeoffs and landings) per hour at these airports is very low. At military airfields, the operations count is on the order of one takeoff or landing per hour. At NASA, there are about three or four launches per year. These statistics are important when discussing CONOPS because what you do depends on how much time you have to do it. With one operation per hour, the controller would have all day to discuss birds on the radio with a pilot.

In contrast, Dallas Fort Worth has 240 operations per hour (i.e., one every 15 seconds). Additionally, the frequencics at these busy airports are so saturated that even if a controller had a bird radar display, he would have no time to discuss the location of the birds with the pilot. What this means for CONOPS is that at military airfields, NASA and low-traffic airports, having a bird radar display in the control tower could be successful if the right operational procedures were employed.

For such low-traffic airports, I would like to offer my opinion on how a bird radar could be employed by air traffic controllers to help pilots avoid collisions with birds.

The CONOPS for the controller will depend on the phase of flight. On takeoff, the controller will see a potential conflict developing and will need to issue an instruction or warning to the pilot. The controller knows the aircraft's planned flight path and will be able to see the bird target flight path. It will require a certain amount of training for the controller to predict where a potential collision will occur. Perhaps the radar software can make a prediction of the bird flight path and create an automated warning. Once a potential conflict is predicted, the controller could issue instructions to fly an alternate heading on takeoff that would avoid the birds. Then, once the aircraft takes off and is clear of the conflict, the controller could vector the aircraft back to course.

On landing approach to the runway, a different procedure could be used. Again, the controller would see a potential collision developing in advance. Several choices could be made. The controller could give the aircraft an avoidance heading to fly and then turn the aircraft back to course. Another possibility would be for the aircraft to do a turn in holding to give the birds time to fly past the potential point of collision. Finally, the aircraft could just go around and climb over the birds to avoid a collision.

For busy airports, the entire process would need to be automated with a collision avoidance system for birds. This would shift the decision making from the controller to the pilot. The CONOPS would depend on the phase of flight. On takeoff, the pilot would observe the programmed flight path and the bird target traffic on an integrated display. When the aircraft is close to takeoff, the pilot would need to determine if a potential conflict existed. If the answer is yes, he could ask the controller for a new heading to avoid the conflict. On approach to landing, the collision avoidance system could provide an automated cockpit warning to alert the pilot of an impending collision. The pilot could either ask the controller for an alternate heading or just push up the power and take the aircraft on another trip around the traffic pattern.

Approve and Fund Airport Purchases of Bird Radars

Only a handful of companies in the world build radars that specifically track and display birds on airport overlays. Most radars can see birds but have always treated them as interference and filtered them out of the final picture that is presented. To develop and integrate a bird detection and display capability into the existing 10- to 40-year-old FAA radar infrastructure, it would take a multibillion-dollar effort, so I don't believe that upgrading these radars to detect and display birds is a good idea anymore. There was a time about 10 years ago when this may have worked, but the concept lacked a following. The fact that NextGen will replace many of the existing ATC radars must be considered. The FAA currently has plans to decommission 125 radars by 2014 as ADS-B is rolled out, a new air transportation system that uses satellite data links and ground transceivers to display airborne targets.

Airports are currently very interested in purchasing bird radars. This is not just because bird radar is like an iPod that they have to own. The reason for ownership is associated with the Federal Aviation Regulations. As a result of the airliner in the Hudson, the FAA is making a new rule that requires all commercial airports (with or without bird problems) to conduct a wildlife hazard assessment (WHA). The FAA has conducted inspections on all commercial airports and has identified approximately 95 airports that are not in compliance. These airports could lose their commercial Part 139 certification because of this violation, which would shut down local commercial airline operations and cause a severe downward impact to the local economy. The airports will most likely want to purchase a bird radar to help with their WHA and wildlife hazard management plan (WHMP) so that their certification is not revoked by the FAA.

The nice thing about an airport management decision to purchase a bird radar is that there is plenty of help available. The federal government will pay for up to 95% of a bird radar once the bird radar evaluation project is complete. Annually, the FAA gives $3.5 billion to airports as part of Airport Improvement Program (AIP) funds. There are 19,500 airports in the United States of which the FAA views 3,411 as critical to our aviation infrastructure. In 2006, only 2,000 of those airports received AIP funds. During the NTSB hearings on Flight 1549, the FAA announced a rulemaking decision to require all 548 commercial airports to conduct a WHA on an initial and periodic basis. So, in the future, this assessment will not only be required by airports with a bird problem but all commercial airports. This requirement, I believe, will drive bird radar sales.

But what about the rest of the 3,411 airports that are critical to our aviation infrastructure? Those airports that apply for and receive AIP funding must comply with 39 grant assurances. In other words, strings are attached to the FAA funds. These grant assurances cover the issues of maintaining a safe operating environment and compatible land use. So, the FAA has regulatory jurisdiction over the 548 commercial airports and can cancel the funding for grant violations at the other 2,000 airports that have received funding if they don't conduct WHAs. The purchase and integration of bird radars at airports should be accomplished through a phased approach. The purchase, operation and maintenance of these radars needs to be well understood. CONOPS need to be developed in order to obtain the greatest benefit from their use. And in parallel, the federal government should fund a collision avoidance system for birds.

Improve Bird Radars

In addition to the technologies explained in this book, bird radars can be improved by integrating other audio and visual recognition systems. A radar will never be able to identify birds by species. Since the methods employed for bird control depends on the species, other reliable methods for species identification should be integrated into bird radars.

Bird species can automatically be classified by the sounds they make. Neural networks, hidden Markov models and dynamic time warping are some methods that have been used to match bird vocalizations with a library to determine a match. Acoustic sensors could be placed around an airport along departure and arrival routes to provide inputs to a signal processing and classification algorithm. Microphone locations could be correlated with bird radar targets to cue the radar to emphasize certain locations where bird targets are present.

Once a bird target is identified by radar, an optical and/or infrared camera could be cued and slewed to the bird target location. Birds could be visually displayed to wildlife control experts to provide a visual means of species identification. Additionally, there are image processing algorithms that will allow the bird species to be automatically identified. Gray level thresholding, color thresholding, RGB thresholding, HSV thresholding, size thresholding, dilation, artificial neural networks and template matching are some of the current methods that should be investigated for automatic classification of birds.

Bird radars could be connected to bird control devices to automate the process of scaring birds away from an airport. Many noisemakers, including pressurized cannons and bird acoustic devices, could be automatically activated depending on the level and location of bird activity on an airport.

Train Pilots on the Hazards of Bird Strikes

Pilots are responsible for reporting all unsafe conditions on or near an airport, including birds that could pose a threat to aircraft safety. When I read about an overseas airline crew that experienced a severe bird strike on takeoff being faulted for the accident, I was shocked. In this case, the crew saw the birds on the runway and even discussed it while they were taxiing out for takeoff. After the takeoff, bird strike, and resulting severe damage to the aircraft, they were faulted for not notifying the tower. The theory was that the tower could have called the wildlife specialists to come out and disperse the birds.

No airline currently trains its crews on the hazards associated with bird strikes. In fact, there is no data to use for the training. I've been flying for over 35 years, and I didn't even know there was an option available to disperse the birds with a call to the tower. Pilots should be trained on how and when to contact wildlife personnel at airports to help with dispersal. If a pilot observes a flock of birds on or near a runway, he should not takeoff or land until the birds are dispersed. Wildlife personnel are usually on call for an immediate response, so the delay should be short. If pilots observe birds that could present a potential strike hazard on the ground or even in flight, they should issue a pilot report (PIREP) so that the controllers can warn other pilots. FAA regulations require controllers to issue advisories that include the bird type, location and direction of flight. These parameters should be included in the PIREP.

If birds are reported along a runway, request a different runway for takeoff and landing. Since 92% of bird strikes occur below 3000 feet, on takeoff, try to climb through this altitude as expeditiously as possible. In flight, consider the speed of the aircraft. If possible, try to fly a slower speed to minimize the impact force of a bird strike. If you encounter birds in flight, be aware that they will dive down when they see the aircraft. A pilot should pull up vs. going down to avoid birds (much like the TCAS logic).

Summary

Once and for all, let's solve this aircraft bird strike problem. Flight 1549 was a wakeup call. More birds and planes take to the skies each year, and if we fail to assemble the pieces of this puzzle, it will just be a matter of time before another airliner goes down from dual engine failure due to multiple bird strikes. We have to stop picking away at this problem with temporary, limited solutions that only delay the inevitable. I've been on technical teams with government and industry, and we have solved some of the most difficult problems involved in the defense of our country, so I know the power of effective collaboration, and I know how achievable these seemingly impossible solutions actually are. I have served with some of the smartest people and the greatest leaders, and I have observed that, ultimately, the greatest problems are solved by people inspired to act selflessly and tackle each challenge that arises. The time for leadership is now. We need a permanent technical solution once and for all. When airliners were colliding with other aircraft in the 80s, and when airliners were colliding with the ground, we analyzed the factors involved, isolated the cause, and developed a solution that we then perfected over time. These were very difficult issues, and our final solution was permanent and well thought-out. When is the last time you remember hearing about a large US airliner hitting another aircraft or hitting the ground? Well, there is your answer, we solved it! As soon as possible, aircraft bird strikes must become a thing of the past.

I close this final chapter by offering my opinion on what needs to be done as a team to bring all of these pieces together into a permanent fix for this problem. This is not just a US problem; it is a worldwide problem, as Appendix 1 clearly shows. Europe and Asia are already moving ahead with plans to purchase bird radars for their airports. Their CONOPS, however, has not been developed yet. We need to form a worldwide organization to standardize the solution so that scarce funds are not wasted on near-term solutions that will most likely just be changed in the future anyway. Aviation is an integral part of the world economy, and it is my deepest hope that teams throughout the world can work together to resolve this problem once and for all.

Appendix 1:

Significant Worldwide Aircraft Bird Strikes for 2009 & History of Serious Aircraft Bird Strike Accidents in the USA

Significant Worldwide Bird Strike Accidents for 2009

(from Italian Bird Strike Consulting & Training Organization)

5 Jan; Bangkok, Thailand, British Airways B744, multiple bird strike on landing;

10 Jan; Firenze, Italy, Alitalia A319, bird strike at takeoff;

15 Jan; New York, NY, US Airways A320, ditched on Hudson river following a multiple bird strike;

16 Jan; Istanbul, Turkey, Lufthansa A321, bird strike on takeoff, one-engine landing;

16 Jan; Guadalajara, Mexico, Interjet A320, bird strike on takeoff, one-engine landing;

17 Jan; Tokyo, Japan, Star Flyer A320, damaged flaps for bird strike on landing;

21 Jan; Ahmedabad, India, Finnair B752, bird strike on landing;

26 Jan; Istanbul, Turkey, Turkish Airlines A321 bird strike on takeoff, one-engine landing;

1 Feb; Teresina, Brazil, GOL B737/800, bird strike at takeoff with ingestion (feral pigeon), back to departing airport;

3 Feb; Denver, CO, United Airlines B757 bird strike at takeoff with ingestion, back to departing airport;

13 Feb; Tacloban, Philippines, Cebu Pacific Airbus 320, bird ingestion at landing, aircraft grounded for maintenance;

16 Feb; S. José, CA, United Airlines B757 rejected takeoff at 120 kts due to a multiple impact with gulls;

25 Feb; Naples, Italy, Air One B737, strike with a single *Larus cachinnans* during final approach, dent in the radome;

6 Mar; Bermuda, American Airlines B 737, bird strike on landing;

8 Mar; Windhoek, Nambia, Air Namibia A340, probable bird strike after takeoff; flight program disrupted for some days;

9 Mar; Tessaloniki, Greece, Easyjet A319, multiple bird strike on takeoff with damages to one engine;

12 Mar; Atlanta, GA, Atlantic Southeast CRJ 200, impact with two geese during climb; precautionary landing with wings and nose damaged;

14 Mar; Shangai, China, JAL B747, bird ingestion after takeoff and immediate diversion on another airport of the city;

16 Mar; New Orleans, LA, Delta B757, bird strike during initial climb and precautionary landing on the departing airport;

17 Mar; Orlando, FL, Southwest B737, impact during the landing roll, minor damages;

18 Mar; Toronto, Canada, Canadair CRJ100, forced to pull up as another airplane on the runway rejected the takeoff clearance due to the presence of birds;

20 Mar; Bandar Lampung, Indonesia, Sriwijaya B737, rejected takeoff for bird ingestion in the left engine;

21 Mar; Bologna, Italy, Eurofly A320, bird strike (2 feral pigeons) during approach, minor damage to a flap;

21 Mar; Newark, NJ, Continental B737, bird strike with ingestion during final approach at about 220 ft.; flame out right engine.

25 Mar; Larnaca, Greece, Aegean A321, multiple bird strike on takeoff and ingestion of 4 birds in the right engine; immediate landing with only one engine on the departing airport;

2 Apr; Luanda, Angola, TAP A340, bird ingestion during initial climb, immediate landing on the departing airport;

4 Apr; Madrid, Spain, Air Nostrum DH8C, bird strike on approach, no damages;

6 Apr; Orlando, FL, Allegiant Air MD83, impact with a bald eagle (*Haliaeetus leucocephalus*) during takeoff, precautionary landing, minor damages;

9 Apr; Sacramento, CA, United Airlines A320, bird strike during initial climb, precautionary landing on departing airport, no damages;

9 Apr; Salvador, Brazil, TAM A320, impact on the left wing during the final approach;

10 Apr; Vancouver, Canada, Air Canada A320, bird ingestion into left engine, rejected takeoff, engine to be changed;

15 Apr; Dallas, TX, American B757, bird strike into right engine on final approach;

17 Apr; Iasi, Romania, Tarom ATR42, collided with a flock of birds on approach, radome damaged;

17 Apr; Atlanta, GA, Atlantic Southeast CRJ2, impact with a large bird on takeoff, precautionary landing, minor damages;

24 Apr; Orlando, FL, Southwest B737, hit two large birds during the landing roll;

5 May; Calgary, Canada, Westjet B737, strike with three Mallard ducks during takeoff roll, struck left engine, precautionary landing. The engine received fan blade;

7 May; Hamilton, New Zealand, Pacific Blue E190, ingestion of plovers in the right engine during takeoff roll, rejected takeoff, three fan blades damaged;

9 May; Mumbai, India, Air India A321, bird ingestion at takeoff, landing on departing airport;

12 May; Amsterdam, Holland, Windjet A320, ingestion at takeoff, landing on departing airport;

14 May; El Salvador, Webjet B737, a vulture (probably an Urubú, *Coragyps atratus*) hit the left wing and got embedded into the wing;

16 May; Konya, Turkey, Turkish A320, bird ingestion after takeoff, return to departing airport;

21 May; Moscow, Russia, Aeroflot A319, ingestion after takeoff, return to departing airport;

23 May; Orlando, FL, Southwest B737, bird strike during short final approach, minor damages.

25 May; Jinan, China, Sichuan Airlines A320, impact at 2000 ft. during approach on the right main gear with damages that forced the plane to do a go-around and an emergency landing;

26 May; Atlanta, GA, Airtran B737, impact during the landing roll;

27 May; Indore, India, Kingfisher A320, bird ingestion into one engine during initial climb, return to the departing airport;

28 May; Fairbanks, AK, ConocoPhillips B737, multiple impact after takeoff, windshield cracked and engine cowling dented, precautionary landing on departing airport;

30 May; San Francisco, CA, Lufthansa A340, bird ingestion into engine #2 on approach;

30 May; Oslo, Norway, SAS B737, a bumble bee (*bombus terrestris*) blocked the pitot system and caused an airspeed indicator malfunction with a precautionary landing on the departing airport;

30 May; Moscow, Russia, Aeroflot B767, struck a flock of birds at takeoff, the flight was continued to Los Angeles where an inspection revealed damage to the right-hand engine;

6 Jun; Darwin, Australia, Jetstar A320, rejected takeoff following a bird strike to the left engine;

7 Jun; Lampedusa, Italy, Windjet A319, bird ingestion into left engine during the approach;

7 June; Cincinnati, OH, American Eagle ERJ 145, collided with a bird on final, landing gear lights damaged;

8 Jun; Vancouver, Canada, Air Canada B777, struck 6 Canada geese during a go-around due to the presence of

birds, relevant damages, including a bird ingestion into the right engine;

10 Jun; Riga, Latvia, Fokker 70, rejected takeoff due to a bird strike on airplane left side;

14 Jun; Diyarbakir, Turkey, Tailwind Airlines B 737, hit from a bird on a wing on final approach, go-around and safe landing;

19 Jun; Bujumbura, Brussels Airline A330, an antelope hit the landing gear during the landing roll, damages;

21 Jun; Montreal, Air Canada ERJ190, birds hit right wing, damage slats and force to an emergency landing;

26 Jun; Chattanooga, TN, Pinnacle CRJ200, struck a number of birds after takeoff, precautionary return on departing airport;

30 Jun; La Guardia, NY, American B737, hit birds on final approach at 900 ft., nose gear damaged with leak of hydraulic fluids;

6 Jul; Ostend, Belgium, World Airways MD-11 ingested a sea gull in one of its engines on takeoff and made an emergency landing;

1 Aug; Parma, Italy, Windjet A319, flew through a flock of gulls ingesting a number of them in the left engine; landed on departing airport with one only engine;

2 Aug; Dhaka, Bangladesh, Livingstone A330, ingested a bird in one engine after the take-off and returned to Dhaka;

2 Aug; Antalya, Turkey, Belair B757, left engine ingested a bird; aircraft returns to Antalya airport;

3 Aug; Derry, Ireland, Ryanair B737, diverted to Dublin after having flown through a flock of birds, airplane under inspection;

5 Aug; Santa Barbara, CA, Mesa Airlines CRJ2, hit a flock of geese at take-off, minor damages;

5 August; Presidente Prudente, Brazil, GOL B 737, ingested a vulture at take off, probably an Urubu;

7 Aug; Saarbrucken, Germany, Luxair DH8D, multiple impact on approach;

8 Aug; London, UK, Norwegian Air Shuttle B737, hit a flock of storks during initial climb, one stork ingested in the left engine;

9 Aug; Raleigh Durham, NC, Northwest A320, hits a bird during take-off run, few minutes later lost left engine and returned to departing airport;

11 August; Glasgow, UK, British Airways A320, bird strike while landing, minor damage to a light glass cover;

11 Aug; Frankfurt, Germany, Lufthansa B747, rejected take-off at high speed due to bird ingestion into an engine;

16 Aug; Fargo, ND, Linx DH8D, hit a flock of birds during climb, returned to Fargo after having burnt fuel for 30';

16 Aug; Paris, France, Cathay Pacific A340; returned to Paris when at FL 310, for a bird ingestion;

17 August; Copenhagen, Denmark, Malev B737, ingested a bird into the right engine, return to CPH; three fan blades damaged;

17 Aug; Toronto, Canada, Air Canada A320, rejected take-off due to an impact with a Red-Tail Hawk (Buteo jamaicensis);

17 Aug; Charlotte, NC, Republic E170, rejected take-off at high speed due to a multiple bird strike with a flock of geese with bird ingestion in the right engine;

22 Aug; Osaka, Japan, ANA B737, a bird entered and struck

the nose gear bay during takeoff damaging a hydraulics tube; the damage forced to a diversion;

23 Aug; Wilmington, NC, Atlantic Southeast CRJ2, rejected take off after a multiple bird strike;

23 Aug; Mumbai, India, Cathay Pacific A 330, struck a bird with the left hand main gear probably during the approach; the strike damaged hydraulic lines;

30 Aug; Calgary, Canada, Air Canada A320, rejected take off after rotation due to a bird strike and stopped safely;

History of Serious Aircraft Bird Strike Accidents in the USA (from Sandy Wright, USDA/WS)

7 Sept 1905; From the Wright Brothers' diaries: "Orville ... flew 4,751 meters in 4 minutes 45 seconds, four complete circles. Twice passed over fence into Beard's cornfield. Chased flock of birds for two rounds and killed one which fell on top of the upper surface and after a time fell off when swinging a sharp curve." This was the first reported aircraft bird strike. Because of the location near Dayton, Ohio and time of year, it's likely that the bird struck was a red-winged blackbird.

3 Apr 1912; Calbraith Rodgers, the first person to fly across the continental USA, was also the first to die as a result of a bird strike. On 3 April 1912, Rodgers' Wright Pusher struck a gull, causing the aircraft to crash into the surf at Long Beach, California. Rodgers was pinned under the wreckage and drowned.

10 Mar 1960; A Lockheed Electra turbo-prop ingested European starlings into all 4 engines during takeoff from Boston Logan Airport (MA). The plane crashed into Boston Harbor, killing 62 people. Following this accident, the FAA initiated action to develop minimum bird ingestion standards for turbine-powered engines.

26 Feb 1973; On departure from Atlanta, Georgia's Peachtree-Dekalb Airport, a Lear 24 jet struck a flock of brown-headed cowbirds attracted to a nearby trash transfer station. Engine failure resulted. The aircraft crashed, killing 8 people and seriously injuring 1 person on the ground. This incident prompted the FAA to develop guidelines concerning the location of solid waste disposal facilities on or near airports.

12 Nov 1975; On departure roll from John F. Kennedy International Airport (NY), the pilot of a DC-10 aborted takeoff after ingesting gulls into 1 engine. The plane ran off runway and caught fire as a result of engine fire and overheated brakes. The resultant fire destroyed the aircraft. All 138 people on board, including airline personnel trained in emergency evacuation, evacuated safely. Following this accident, the National Transportation Safety Board recommended the FAA evaluate the effect of bird ingestion on large, high-bypass, turbofan engines and the adequacy of engine certification standards. The FAA initiated a nationwide data collection effort to document bird strike and engine ingestion events.

25 July 1978; A Convair 580 departing Kalamazoo Airport (MI) ingested 1 American kestrel into an engine on takeoff. Aircraft auto-feathered and crashed in nearby field, injuring 3 of 43 passengers.

18 June 1983; The pilot of a Bellanca 1730, landing at Clifford TX, saw 2 "buzzards" on final approach. He added power and maneuvered to avoid them, then continued approach. This resulted in landing beyond intended point. The middle of runway was higher than either end; therefore, pilot was unable to see a large canine moving toward the landing area until aircraft was halfway down runway. A go-around was initiated but the lowered landing gear hit some treetops causing the pilot to loose control. The aircraft came to rest in a milo field about 250 yards from initial tree impact after flying through additional trees. Aircraft suffered substantial damage, and 2 people in aircraft were seriously injured.

5 Nov 1990; During takeoff at Michiana Regional Airport (IN), a BA-31 flew through a flock of mourning doves. Several birds were ingested in both engines and takeoff was aborted. Both engines were destroyed. Cost of repairs was $1 million and time out of service was 60 hours.

30 Dec 1991; A Citation 550, taking off from Angelina County Airport (TX) struck a turkey vulture. The strike caused major damage to #1 engine and resulting shrapnel caused minor damage to the wing and fuselage. Cost of repairs was $550,000 and time out of service was 2 weeks.

3 Dec 1993; A Cessna 550 struck a flock of geese during initial climb out of DuPage County Airport (IL). Pilot heard a loud bang and aircraft yawed to left and right. Instruments showed loss of power to #2 engine and a substantial fuel leak on the left side. An emergency was declared, and the aircraft landed at Midway Airport. Cost to repair 2 engines was $800,000, and time out of service was about 3 months.

3 June 1995; An Air France Concorde, at about 10 feet AGL while landing at John F. Kennedy International Airport (NY), ingested 1 or 2 Canada geese into the #3 engine. The engine suffered an uncontained failure. Shrapnel from the #3 engine destroyed the #4 engine and cut several hydraulic lines and control cables. The pilot was able to land the plane safely, but the runway was closed for several hours. Damage to the Concorde was estimated at over $7 million. The French Aviation Authority sued the Port Authority of New York and New Jersey and eventually settled out of court for $5.3 million.

5 Oct 1996; A Boeing-727 departing Washington Reagan National Airport (DC) struck a flock of gulls just after takeoff, ingesting at least 1 bird. One engine began to vibrate and was shut down. A burning smell entered the cockpit. An emergency was declared, and the aircraft, carrying 52 passengers, landed at Washington National. Several engine blades were damaged.

7 Jan 1997; An MD-80 aircraft struck over 400 blackbirds just after takeoff from Dallas-Fort Worth International Airport (TX). Almost every part of the plane was hit. Pilot declared an emergency and returned to land without event. Substantial damage was found on various parts of the aircraft and the #1 engine had to be replaced. The runway was closed for 1 hour. The birds had been attracted to an unharvested wheat field on the airport.

9 Jan 1998; While climbing through 3000 feet, following takeoff from Houston Intercontinental Airport (TX), a Boeing-727 struck a flock of snow geese with 3-5 birds ingested into 1 engine. The engine lost all power and was destroyed. The radome was torn from aircraft and leading edges of both wings were damaged. The pitot tube for first officer was torn off. Intense vibration was experienced in airframe and noise level in cockpit increased to point that communication among crewmembers became difficult. An emergency was declared. The flight returned safely to Houston with major damage to aircraft.

22 Feb 1999; A Boeing-757 departing Cincinnati/Northern Kentucky International Airport (KY) had to return and make emergency landing after hitting large flock of starlings. Both engines and 1 wing received extensive damage. About 400 dead starlings were found on runway area.

21 Jan 2001; The #3 engine on an MD-11 departing Portland International Airport (OR) ingested a herring gull during takeoff run. The bird ingestion resulted in a fractured fan blade. Damage from the fan blade fracture resulted in the liberation of the forward section of the inlet cowl. Portions of the inlet cowl were ingested back into the engine and shredded. The pilot aborted takeoff during which two tires failed. The 217 passengers were safely deplaned and rerouted to other flights. Bird ID by Smithsonian, Division of Birds.

09 Mar 2002; A Canadair RJ 200 at Dulles International Airport (VA) struck 2 wild turkeys during the takeoff

roll. One shattered the windshield spraying the cockpit with glass fragments and remains. Another hit the fuselage and was ingested. There was a 14" x 4" section of fuselage skin damaged below the windshield seal on the flight officer's side. Cost of repairs estimated at $200,000. Time out of service was at least 2 weeks.

19 Oct 2002; A Boeing 767 departing Logan International Airport (MA) encountered a flock of over 20 double-crested cormorants. At least 1 cormorant was ingested into #2 engine. There were immediate indications of engine surging followed by compression stall and smoke from engine. The engine was shutdown. Overweight landing with 1 engine was made without incident. Nose cowl was dented and punctured. There was significant fan blade damage with abnormal engine vibration. One fan blade was found on the runway. Aircraft was towed to the ramp. Hydraulic lines were leaking, and several bolts were sheared off inside engine. Many pieces fell out when the cowling was opened. Aircraft was out of service for 3 days. Cost of repairs was $1.7 million.

8 Jan 2003; A Bombardier de Havilland Dash 8 collided with a flock of lesser scaup at 1,300 feet AGL on approach to Rogue Valley International Airport (OR). At least 1 bird penetrated the cabin and hit the pilot who turned control over to the first officer for landing. Emergency power switched on when the birds penetrated the radome and damaged the DC power system and instruments systems. The pilot was treated for cuts and released from the hospital.

04 Sept 2003; A Fokker 100 struck a flock of at least 5 Canada geese over runway shortly after takeoff at LaGuardia Airport (NY), ingesting 1 or 2 geese into #2 engine. Engine vibration occurred. Pilot was unable to shut engine down with the fuel cutoff lever, so fire handle was pulled and engine finally shut down, but vibration continued. The flight was diverted to nearby JFK International Airport where a landing was made. The NTSB found a 20" x 36" wide depression on right side of nose behind radome. Maximum depth was 4 inches.

Impact marks on right wing. A fan blade separated from the disk and penetrated the fuselage. Several fan blades were deformed. Holes were found in the engine cowling. Remains were recovered and identified by Wildlife Services.

17 Feb 2004; A Boeing 757 during takeoff run from Portland International Airport (OR) hit 5 mallards and returned with 1 engine out. At least 1 bird was ingested and parts of 5 birds were collected from the runway. Engine damage was not repairable and engine had to be replaced. Cost was $2.5 million and time out of service was 3 days.

15 Apr 2004; An Airbus 319 climbing out of Portland International Airport (OR) ingested a great blue heron into the #2 engine, causing extensive damage. Pilot shut the engine down as a precaution and made an emergency landing. Runway was closed 38 minutes for cleaning. Flight was cancelled. Engine and nose cowl were replaced. Time out of service was 72 hours. Damage totaled $388,000.

14 June 2004; A Boeing 737 struck a great horned owl during a nighttime landing roll at Greater Pittsburgh International Airport (PA). The bird severed a cable in front main gear. The steering failed, the aircraft ran off the runway and became stuck in mud. Passengers were bused to the terminal. They replaced 2 nose wheels, 2 main wheels and brakes. Aircraft out of service was 24 hours. Cost estimated at $20,000.

16 Sept 2004; A MD 80 departing Chicago O'Hare (IL) hit several double-crested cormorants at 3000 feet AGL and 4 miles from airport. The #1 engine caught fire and failed, sending metal debris to the ground in a Chicago neighborhood. The aircraft made an emergency landing back at O'Hare with no injuries to the 107 passengers.

24 Oct 2004; A Boeing 767 departing Chicago O'Hare (IL) hit a flock of birds during takeoff run. A compressor stall caused the engine to flame out. A fire department got calls from local residents who reported seeing flames coming from

the plane. Pilot dumped approximately 11,000 gallons of fuel over Lake Michigan before returning to land. Feathers found in engine were sent to the Smithsonian, Division of Birds for identification.

30 Mar 2005; A SA 227, landing at Dade-Collier Training and Transportation Airport (FL), hit the last deer in a group of 8 crossing the runway, causing a prop to detach and puncture the fuselage. Also damaged was the nose wheel steering and right engine nacelle. Aircraft was a write-off due to cost of repairs $580,000 being close to the plane's value of $650,000.

1 Sept 2005; A Falcon 20 departing Lorain County (OH) Airport hit a flock of mourning doves at rotation, causing the #1 engine to flame out. As the gear was retracted, the aircraft hit another flock, which caused the #2 engine RPM to roll back. The pilot was not able to sustain airspeed or altitude and crash-landed, sliding through a ditch and airport perimeter fence, crossing a highway and ending in a corn field. Aircraft sustained major structural damage beyond economical repairs. Both pilots were taken to hospital. Costs totaled $1.4 million.

16 Oct 2005; A BE-1900 departing Ogdensburg International (NY) struck a coyote during takeoff run. The nose gear collapsed causing the plane to skid to a stop on the runway. Propeller blades went through the skin of the aircraft. Engine #1 and #2, propellers, landing gear, nose, fuselage had major damage. Insurance declared aircraft a total loss. Cost of repairs would have been $1.5 million.

30 Dec 2005; A pilot flying a Bell 206 helicopter at 500 feet AGL near Washington, LA looked up from instruments to see a large vulture crashing into the windshield. He was temporarily blinded by blood and wind. After regaining control, the pilot tried to land in a bean field nearby but blood was hampering his vision and the left skid hit the ground first causing the aircraft to tip on its side. Pilot was taken to

the hospital and had several surgeries to repair his face, teeth and eye. Aircraft was damaged beyond repair. Cost of repairs would have been $1.5 million.

1 Jan 2006; A B-757 ingested a great blue heron into an engine during takeoff at Portland International (OR). Engine was shut down and a one-engine landing was made. Fan section of the engine was replaced. Time out of service was 15 hours. Cost was $244,000.

3 Aug 2006; A Cessna Citation 560 departing a General Aviation airport in Indiana hit Canada geese on the takeoff run. Left engine ingested birds causing an uncontained failure. Aircraft went off the runway during the aborted takeoff. Top cowling and fan were replaced. ID by the Smithsonian, Division of Birds. Aircraft was out of service for 13 days and costs were estimated at $750,000.

18 Aug 2006; A CL-RJ 200 departing Salt Lake City International Airport flew through a flock of northern pintails (ducks) at 500 feet AGL. Pilot saw 2 birds and felt them hit the engines. Engines began to vibrate. Aircraft landed without incident and was towed to the hangar. ID by the Smithsonian, Division of Birds. Time out of service was over 24 hours and costs to repair engines totaled $811,825.

8 Dec 2006; The Captain of a B-767 departing JFK International Airport saw 2 birds during initial climb. After bird was ingested into #2 engine, pilot returned aircraft to JFK on Alert 3-3. One badly damaged great blue heron was recovered from the runway. Carcass appeared to have gone through the #2 engine. The engine was replaced and passengers were put on a replacement aircraft.

15 Mar 2007; A B-767 departing Chicago O'Hare encountered a flock of birds at <500 feet AGL. People on ground reported flames shooting out of the #1 engine. The aircraft returned to land without incident and was towed to the terminal. Birds were ingested in both engines, but only

1 engine was damaged. Remains of nine male canvasback ducks were found near the departure end of runway 9R. ID by the Smithsonian, Division of Birds. Time out of service was 12 days. Estimated cost for repairs is $1.8 million. Cost for aircraft's time out of service was $309,000.

7 July 2007; A US carrier B-767 flew through a large flock of yellow-legged gulls at 20 feet AGL during departure at Fiumcino International Airport (Rome, Italy). The pilot dumped fuel before returning to land on one engine. Besides birds being ingested into both engines, birds hit the cockpit window, right engine nose cowl, wing, and right main undercarriage. The main gear struts were deflated. Some of the fan blades had large chunks taken out. The left engine had many fan blades damaged midway along the blade leading edge. Both engines were replaced. The replacement engines had to be flown to Rome from the USA. ID by ornithologist, a member of Bird Strike Committee Italy. Time out of service 1 week.

25 Aug 2007; Pilot of B-737 departing Texas El Paso Airport reported loud bang in cockpit at 14000 feet AGL during climb. Loud rushing air noise, cabin started to depressurize. Cabin altitude horn went off, oxygen masks were donned. Pilot descended to 10000 feet, notified flight attendants of situation, and then landed at El Paso. Found large hole under captain's left foot side. Also, hole in left horizontal stabilizer the size of a football. First officer's side of cockpit had a dent. Blood and feathers were found. No birds were seen in flight. Ground crew said "turkey buzzards" were in area. Bird was identified as marbled godwit by Smithsonian, Division of Birds. Cost of repairs was $144,064. Time out of service was 3 days.

28 Aug 2007; The pilot of a CRJ-700 declared an emergency after a black vulture smashed in the front fuselage between the radome and the windshield at 2,300 feet AGL on approach to the Louisville, KY International Airport. The strike ripped the skin, broke the avionics door, broke a stringer

in half and bent 2 bulkheads. Maintenance made temporary repairs, then aircraft was ferried out for permanent repairs. ID by Smithsonian, Division of Birds. Cost of repairs was $200,000. Time out of service was 2 weeks.

11 Oct 2007; A CRJ-700 departing Denver International struck a flock of sandhill cranes at 1,500 feet AGL. The captain said several "geese" came at them, and they heard 3-4 thuds. The right engine immediately began to run roughly, and the VIB gauge was fluctuating rapidly from one extreme to the other. Captain declared an emergency and said he didn't think he was going to make it back to DEN. The aircraft landed safely. The engine fan was damaged, and there were dents along the left wing leading edge slat. ID by Smithsonian, Division of Birds. NTSB investigated.

23 Oct 2007; A Piper 44 flying at 3,400 feet AGL disappeared during a night training flight from Minneapolis, MN to Grand Forks, ND. The instructor and student pilot did not report any difficulties or anomalies prior to the accident. Wreckage was found 36 hours later, partially submerged upside down in a bog. The NTSB sent part of a wing with suspected bird remains inside to the Smithsonian. Remains identified as Canada goose. The damage that crippled the aircraft was to the left horizontal stabilator. NTSB investigated. Two fatalities.

22 Nov 2007; Pilot of a B-767 (US carrier) at Nice Cote d'Azur (France) noticed a flock of gulls on runway during takeoff. As the aircraft rotated, the flock lifted off the runway. Shortly after that, the crew felt multiple strikes and vibrations and returned to land. The #2 engine had fan blade damage. One piece of a fan blade broke off and exited out the front and the core nozzle fell off. The engine was replaced. Birds were identified as yellow-legged gulls by Smithsonian, Division of Birds. Time out of service was 12 days. Cost of repairs was $8,925,000 and other cost was $196,000.

27 Nov 2007; A CRJ-200 descending into Memphis

International Airport (TN) encountered a flock of large birds, sustaining ingestion into both engines, a cracked nose panel, damage to the right wing root and left horizontal stabilizer, and left engine anti-ice cowling. Bird remains were subsequently identified as snow geese. Maintenance made temporary repairs before aircraft could be flown for more permanent repairs.

29 Jan 2008; Flight crew of B-747 reported minor noise and vibration shortly after lift-off from Louisville International Airport. Noise and vibrations later subsided. Upon landing at destination, damage was found to 3 fan blades on the #2 engine. A piece of a liberated fan blade penetrated the cowl. Six fan blade pairs, the fan case outer-front acoustic panel and inlet cowl were replaced. ID by Smithsonian, Division of Birds.

8 Apr 2008; Shortly after departure, a Challenger 600 suffered multiple, large bird strikes (American white pelicans) at 3000 feet AGL. One bird penetrated the nose area just below the windshield and continued through the forward cockpit bulkhead. Bird remains were sprayed throughout the cockpit. No injuries reported. Both engines ingested at least 1 bird. The #1 engine had fan damage; the #2 engine lost power and had a dented inlet lip. ID by Smithsonian, Division of Birds. NTSB investigated. Cost exceeded $2 million.

20 Jun 2008; During takeoff run at Chicago O'Hare, a B-747 bound for China ingested a red-tailed hawk. The flight continued takeoff and climbed to 11000 feet to dump fuel and then returned to the airport with one engine out. Several blades had significant damage. Both the #1 and #2 engines had vibrations, but the #2 engine was not damaged. Aircraft taken out of service for repairs; passengers had to be boarded onto another aircraft.

Large Military Aircraft

15 July 1996; Belgian Air Force Lockheed C-130;

Eindhoven, Netherlands; The aircraft struck a flock of birds during approach and crashed short of the runway. All four crew members and 30 of the 37 passengers were killed.

14 July 1996; NATO E-3 AWACS; Aktion, Greece; The aircraft struck a flock of birds during takeoff. The crew aborted the takeoff, and the aircraft overran the runway. The aircraft was not repaired, but none of the crew was seriously injured.

22 Sept 1995; US Air Force E-3 AWACS; Elmendorf AFB, Alaska; During takeoff as the aircraft was passing rotation speed, the aircraft struck about three dozen geese, ingesting at least three into engine two and at least one into engine one. The aircraft was unable to maintain controlled flight and crashed in a forest about 1 mile (1.6 km) beyond the runway. All 24 occupants were killed.

Sept 1987; US Air Force B1-B; Colorado, USA; Aircraft lost control and crashed after a large bird struck the wing root area and damaged a hydraulic system. The aircraft was on a low-level, high-speed training mission. Only three of the six occupants were able to successfully bail out.

1980; Royal Air Force Nimrod; Kinloss Scotland; Aircraft lost control and crashed after ingesting a number of birds into multiple engines.

APPENDIX 2:

Turbine Engine Foreign Object Ingestion and Rotor Blade Containment Type Certification Procedures

(From Advisory Circular AC 33-1B dated April 22, 1970)

Background: Experience acquired with turbine engines has revealed that foreign object ingestion has, at times, resulted in safety hazards. Such hazards may be extreme and possibly catastrophic, involving explosions, uncontrollable fires, engine disintegration, and lack of containment of broken blading. In addition, lesser but potentially severe hazards may involve airflow disruption with flameouts, lengthy or severe power losses, or momentary disruptions and possibly minor blade damage. While the magnitude of the overall hazards from foreign object ingestion is often dependent upon more than one factor, engine design appears to be the most important.

Scope: For the purpose of showing compliance with the reference regulations, engine type certification programs should include substantiation of engine ingestion properties and broken rotor blade damage containment. To insure the provision of a desired degree of engine tolerance to the disruptive effects of foreign object ingestion, substantiation should include an evaluation of the engine design and tests to demonstrate the ability to ingest typical foreign objects without causing a serious reduction in flight safety. The engine applicant is permitted to specify the use of protected inlets for his engine as an alternative to substantiation for airborne foreign objects.

Desirable Engine Design Features: Experience has indicated that the following design features have generally minimized severe effects of foreign object ingestion and effectively increased safety:

a. Front rotor blades, inlet and stator vanes material and design which minimize impact damage, severe deflections, tearing, and rupture.

b. Shrouded tips for the first several rotor and stator stages.

c. Front stage compressors (or fans) without entry guide vanes.

d. Blades which effectively mince birds upon contact.

e. Appreciable axial clearance between the first-stage compressor rotor blades and entry guide vanes and between rotor blades and stators of the first several compressor stages, especially near the rotor blade forward tips.

f. Puncture- and tear-resistant rotor housings or separate armor adequate to contain broken rotor blades and stator vanes.

g. Adequate strength of engine main structure and bearing supports to provide a strength margin to permit safe shutdown and low-speed windmilling when large unbalances typical of damaged rotor blading occur.

h. A generous stall margin for the engine, good combustion stability during airflow disturbances incident to foreign object ingestions, and rapid relight capability.

Classification of Foreign Objects: For the engine substantiation program, the foreign objects considered typical are classified into two major groups.

A. Foreign objects in Group I are those applicable to all turbine engines and are likely to be encountered only as single occurrences affecting just one engine of any multi-engine aircraft in any one flight.

Group I:

(1) A cleaning cloth of typical size;

(2) A mechanic's hand tool of pocket size;

(3) A small-size aircraft steel bolt and nut typical of aircraft inlet hardware;

(4) A piece of aircraft tire tread of length equal to the tread width of a representative size tire;

(5) Compressor and turbine rotor blades: The most critical single blade(s), usually of the largest size, with failure assumed in or adjacent to the outermost retention member. While the majority of failures are expected to occur in the blade airfoil section, failures in or near the retention sections of the blade are also anticipated and are more difficult to contain in the engine. For integrally bladed rotors, failure of a significant portion of a blade should be assumed. While rotor blades are not normally to be categorized as foreign objects in their respective engines, failed blades are so considered for the purpose of this circular; and

(6) Birds of 4 lbs. and over (geese, buzzards, largest gulls, ducks).

B. Foreign objects in Group II are those considered to be generally airborne as regards their reason for entry into engines and may be ingested by more than one engine of an aircraft on any one occasion. Since all engines of an aircraft, whether single or multi-engine, may be affected by ingestions in the same flight, power recovery level is covered herein. Unless the specific installation, inlet design, or other factors preclude the possibility of the ingestion of particular foreign objects, all of the following objects are applicable:

Group II:

(1) Water in the form of rain and snow;

(2) Gravel of mixed sizes up to one-fourth inch typical of airport surface material in quantities likely to be ingested in one flight;

(3) Sand of mixed sizes typical of airport surface material in quantities likely to be ingested in several flights;

(4) Ice of typical sizes and forms representative of inlet duct and lip formations, engine front frame and guide vane deposits, in quantities likely to be ingested during a flight;

(5) Hail stones of approximately 0.8 to 0.9 specific gravity and of one- and two-inch diameter; and

(6) Birds in weight categories as follows:

(a) Small birds of 2 to 4 ounces (starlings);

(b) Medium birds of 1 to 2 pounds (common gulls, small ducks, pigeons).

Substantiation Tests:
A. Group I Foreign Objects:

(1) Ingestion Group I foreign objects except rotor blades and large birds while operating at maximum output. The typical objects being ingestion tested are normally introduced by dropping them into the inlet. It is desired that engine operation after ingestion be continued to determine whether the engine is in a condition of imminent failure, particularly when some unbalance is present.

(2) Ingestion of broken rotor blades: Rotor blades are to be

evaluated for both ingestion effects and containment, and should be released from a rotor at maximum operating rpm excluding transient overspeeds. The rotor blades evaluated normally include all those which, in combination with the adjacent rotor case wall section, are likely to be the most difficult to contain. If the engine continues to operate, observe a representative delay of about 15 seconds in initiating engine shutdown after the first indication of a fault from engine instruments following blade ingestion, to simulate crew reaction time and determine the short-term effects of operation with this unbalance. Longer post-ingestion operation should be accomplished to determine the effects of questionable internal damage, which may not be readily indicated by engine instruments.

(3) Ingestion of a 4-lb. bird: Ingest one large bird of at least 4 lbs. weight if it can enter the inlet and reach the engine face. Ingest at typical maximum climb speed with takeoff power output.

B. Group II Foreign Objects:

(1) General

(a) The engine front face, including the nose cone area, should be tested to substantiate direct impact effects. This may be accomplished as component tests;

(b) Damage resulting from ingesting airborne foreign objects could cause blade damage or failures, and tolerance to this should be evaluated with an operating engine.

(c) The provision of a wind tunnel facility to provide a moving airstream into the test engine is desirable, but is not essential where the injection of the foreign objects into the operating engine to simulate the effects of aircraft speed is adequate. Whenever results considered particularly critical to safety result from ingestion tests, however, it is desirable to conduct either a wind tunnel test, a flight test, or a particularly accurate simulation of flight effects on the severity of ingestion effects. As an example, the minimum propeller blade pitch settings used with turbopropeller engines in flight may require special test settings under static test stand conditions to simulate flight operation characteristics.

(d) Bird sizes, weights, and quantities indicated for test purposes are based on ingestion experience. There are numerous instances reported of small bird and medium bird ingestions, both singly and in multiples. Large birds have been encountered singly in all but a very few instances. Ingestion testing the large bird is aimed primarily at substantiating direct impact effects. Ingestion testing of the smaller birds is aimed at substantiating the effects of bird masses. Both inlet opening width and overall area have a bearing on the probability of ingestion of given size birds and these factors, along with bird flocking density, were considered in selecting bird sizes and quantities.

(e) Duration of the engine running following ingestion of any Group II objects should be at least five minutes to determine whether the engine is in a condition of imminent failure, but, in case of doubt as to actual engine condition or evident engine damage, longer post-ingestion test runs should be conducted.

(2) Hail Ingestion

(a) Hail ingestion should be tested with operating engines. To simulate the effects of random strikes on different portions of the engine face, hailstones should be aimed at areas which could be the most critical for hail impact including the nose fairing.

(b) With single 1-inch hailstones, ingest at the maximum cruising flight speed representative of the applicable type aircraft with maximum cruise engine output.

(c) With single 2-inch hailstones, ingest at climb (or

rough air) speed for up to 15000 feet altitude, whichever is the highest speed with maximum continuous engine output. The tests with 2-inch hailstones may be omitted if the kinetic energy developed with the 1-inch hail tests is as high.

(d) For multiple hailstone testing, ingest the following test quantities sequenced at close intervals to simulate random encounters at climb (or rough air) speed and with maximum continuous power level. Test quantities suggested for each 150 square inches or less of inlet area are either one l-inch and one 2-inch hailstone or, if l-inch hail is used in accordance with foregoing paragraph (c), use two l-inch hailstones. For engines of 100 square inches inlet area or less, only one l-inch hailstone is applicable.

(3) Ice Ingestion: Ice should be introduced into an engine operating at cruise conditions to simulate typical ice shedding from inlets and the engine front face because of possible delays in operating ice protection systems. Engine flameout tendencies, re-ignition, and power recovery capability should be evaluated.

(4) Water Ingestion: An acceptable water ingestion testing method simulates maximum rainfall in quantities up to approximately four percent of the engine weight airflow with engine operating at stabilized cruise and sea level takeoff power levels for at least 3 minutes.

(5) Bird Ingestion: Bird ingestion tests using freshly killed birds and gun injection are preferable as actual strikes are closely simulated. Other acceptable techniques have been used which utilize previously frozen birds and injection means other than guns. If previously frozen birds are used, they should be completely thawed for the tests, and have normal moisture content. If frozen for appreciable periods, moisture content may be reduced below normal levels. Use of synthetic "birds" has been proposed and will be acceptable if the results of ingestion can be shown to be equivalent to

ingesting actual birds. For testing impact effects, include all frontal areas considered to be critical and appropriate bird velocities indicated in the following paragraphs should be attained at the inlet. Other ingestion effects, such as compressor stall or blowout, may be sufficiently severe at somewhat lower bird velocities.

(a) Small Birds: Ingest at typical takeoff flight speeds and engine output levels. Ingest one small bird for each 50 square inches of inlet area (or fraction thereof) if it can enter the inlet and reach the engine face. The maximum number of small birds to be ingested as a group need not exceed 16. Small bird testing may be omitted for large engines when it is agreed that medium birds will pass into the engine blading passages and result in a test of at least equal severity.

(b) Medium Birds: Ingest at typical initial climb speed with takeoff engine output. When medium birds can enter the inlet and reach the engine intake section, ingest one bird for each 300 square inches (or fraction thereof) of intake area up to 3,000 square inches, with additional medium birds at 1/3 of this rate for larger engines. Small and medium birds should be ingested in random sequence, dispersed over the inlet area, to simulate an encounter with a flock.

Substantiation Criteria

A. Rotor Blade Containment: The engine is acceptable if, during the tests, the damage from rotor blade failures is contained by the engine, e.g., without causing significant rupture or hazardous distortion of the engine casing and the expulsion of blades through or beyond the edge of the engine case or shield.

B. Ingestion Hazards:

(1) Group I Objects: The engine is acceptable if ingestion tests demonstrate freedom from engine explosion, disintegration, or uncontrollable fire. It is acceptable that the engine may

require shutdown, but this should be indicated by excessive vibration or other direct operating evidence in a timely manner which would permit a safe shutdown.

(2) Group II Objects: The engine is acceptable if tests demonstrate freedom from the foregoing hazards and the ability to minimize overall hazards and potentially serious conditions, with the quantities and conditions indicated, by its continued safe operation after the ingestion tests. There should be no indication of need for immediate shutdown or imminent failure during the ingestion tests, and prompt engine recovery should be obtained. There should be no flameouts or significant sustained power loss from ice, hail, or water ingestion or hazardous effects from case contraction from the water ingestion test. Power recovery to stabilized operation following other Group II ingestions may be at reduced levels, and the desired minimum level is 75 percent.

C. Limitations: If an engine has not met the criteria in B above, or when the use of protective inlets is elected by the applicant in lieu of ingestion testing Group 2 foreign objects, then the engine should be used only with acceptable inlet protection. The engine manufacturer's installation data and the engine type certificate sheet should, therefore, indicate that a protective inlet is needed when none is incorporated in the engine. The qualification of specific "protective" inlets for both normal function in the aircraft and ability to effectively exclude specified foreign objects is normally a part of aircraft certification.

Appendix 3:

Bird Ingestion Certification Standards for Jet Engines

(From 14 CFR Part 33 October 17, 2007)

Summary: This final rule amends the aircraft turbine engine type certification standards to better address the threat flocking birds present to turbine engine aircraft. These changes will also harmonize FAA and European Aviation Safety Agency (EASA) bird ingestion standards for aircraft turbine engines certified by the United States and the EASA countries, and simplify airworthiness approvals for import and export. The changes are necessary to establish uniform international standards and provide an acceptable level of safety for aircraft turbine engines with respect to the current large flocking bird threat.

Background

The FAA adopted new regulations under 14 CFR 33.76 on September 5, 2000 to better address the overall bird ingestion threat to turbine powered aircraft. These requirements were adopted, in part, as a response to NTSB safety recommendation A-76-64, which recommended an increase in the level of bird ingestion capability for aircraft engines.

Based on comments received during that rulemaking effort, the FAA decided to pursue additional rulemaking to address larger flocking birds (mass greater than 1.15 kg/2.5 pounds), since existing engine certification requirements did not specifically address the threat that these size birds, or their growing population, present to airplane operational safety.

Summary of the NPRM

On July 20, 2006, the FAA published a notice of proposed rulemaking (NPRM), "Airworthiness Standards; Engine Bird Ingestion" (71 FR 41184). The NPRM proposed to amend aircraft turbine engine type certification standards to reflect recent analysis of the threat flocking birds present to turbine engine aircraft. The proposed changes are necessary to establish uniform international standards that provide an adequate level of safety. The comment period closed September 18, 2006.

Summary of the Final Rule

The final rule adopts new bird ingestion standards for turbine aircraft engines under 14 CFR 33.76. It also provides a detailed description of the rulemaking project, including the safety objective and a discussion of the considerations supporting our selection of this course of action. No changes were made to the final rule from what was proposed in the NPRM.

Summary of Comments

The FAA received comments from Transport Canada Civil Aviation (TCCA) and the National Transportation Safety Board (NTSB). TCCA fully supports the intent of the proposal. However, NTSB expressed concern with the size of the largest bird upon which the rule is based (8 lbs). NTSB reasoned that flocking birds greater than 8 pounds can exist in the environment, and may have impacted commercial aircraft in the past. NTSB also expressed concern about using de-rated takeoff thrust instead of full-rated takeoff thrust value for required tests because full-rated thrust can be selected by the flight crew, and because this power setting may be a more severe case than using de-rated takeoff thrust. NTSB suggested the required tests be revised to reflect a worst-case scenario.

The FAA does not concur with these three comments. The safety objective of this rule is to address the expected world fleet rate of catastrophic aircraft events due to multi-engine power loss resulting from multi-engine ingestion of large flocking birds. The various rule parameters were carefully selected to achieve this goal by devising tests that encompass a sufficient percentage of possible parameter combinations (e.g., bird mass/number, bird speed, engine power setting, target locations, etc.) that would allow the world fleet to operate at this very high level of safety. The database of ingestion events used to determine ingestion rates covers a 30-year period and over 325 million flights. The database analysis enabled the FAA to define the actual threat experienced in service, including a conservative adjustment for potential future increases in ingestion rates. The proposed rule was not intended to encompass the worst possible combination of factors, as this is problematic to predict, and would be beyond the capability of current engine technology. We believe selecting all parameters using a theoretical worst-case scenario would be impractical from a design, manufacture, and operational standpoint.

NTSB further suggested incorporating pre-existing fan blade service damage into the required tests because the potential exists for such damage to occur in normal service. The FAA is not adopting this suggestion. Engine type certification requirements are intended for, and applied to, undamaged products as a baseline. The engine bird ingestion requirements and type certificate (TC) requirements are similar in this regard. This revised rule is based on critical ingestion parameters for the most severe engine bird ingestion events recorded over the past several decades. As such, substantial margin exists for the normal ingestion events seen in service, including service acceptable damage allowed by the Instructions for Continued Airworthiness (ICAs). Also, current Advisory Circular material for ICA compliance specifies the type certificate holder evaluate service-acceptable damage criteria against the type certification requirements, and include appropriate instructions in the ICAs. The overall positive experience of the world fleet indicates that this general approach provides an acceptable level of safety.

NTSB also suggested that the FAA consider bird ingestion event data collected since the bird study cutoff date of 1999. NTSB asserts the 30-year data set used is inadequate to assess the risk associated with bird ingestion. The FAA's decision to proceed with this rulemaking is based on quantitative and qualitative evaluation of the threat observed in service over a lengthy period of time. We concluded that the increasing population of large flocking birds in the environment, and the increasing number of encounters in service, make it necessary to expand the scope of the existing requirements. The data from the 30-year study period covers over 325 million flights and is comprised of data from actual engine bird ingestion events where the bird species, size, and number; aircraft and engine model; flight regime; and outcome are reasonably known. The database covers a broad cross-section of aircraft type and operations and is considered fully adequate to establish engine bird ingestion rates from which the critical ingestion parameters were selected to meet the rule's safety objective. The event data collected since the study period does not appear to indicate a change in the basic threat definition or an increase in the actual rate of occurrence and would not likely affect the outcome of the rulemaking project. Finally, as suggested by Transport Canada Civil Aviation (TCCA), the FAA has reviewed the new table included in the amendatory language to ensure it is accurate. The final rule is adopted as proposed.

The Amendment

In consideration of the foregoing, the Federal Aviation Administration amends Chapter I of Title 14, Code of Federal Regulations as follows:

PART 33—AIRWORTHINESS STANDARDS: AIRCRAFT ENGINES

1. The authority citation for part 33 continues to read as follows:

Authority: 49 U.S.C. 106(g), 40113, 44701, 44702, 44704.

2. Amend Sec. 33.76 by revising paragraphs (a) introductory text, (a)(1), (a)(3), (a)(5), the heading of paragraph (b) introductory text, and the heading of paragraph (c) introductory text, and adding paragraph (d) and Table 4 to read as follows:

Sec. 33.76 Bird ingestion.

(a) General. Compliance with paragraphs (b), (c), and (d) of this section shall be in accordance with the following:

(1) Except as specified in paragraph (d) of this section, all ingestion tests must be conducted with the engine stabilized at no less than 100-percent takeoff power or thrust, for test day ambient conditions prior to the ingestion. In addition, the demonstration of compliance must account for engine operation at sea-level takeoff conditions on the hottest day that a minimum engine can achieve maximum rated takeoff thrust or power.

(2) The impact to the front of the engine from the large single bird, the single largest medium bird which can enter the inlet, and the large flocking bird must be evaluated. Applicants must show that the associated components when struck under the conditions prescribed in paragraphs (b), (c) or (d) of this section, as applicable, will not affect the engine to the extent that the engine cannot comply with the requirements of paragraphs (b)(3), (c)(6) and (d)(4) of this section.

(3) Objects that are accepted by the Administrator may be substituted for birds when conducting the bird ingestion tests required by paragraphs (b), (c) and (d) of this section.

(b) Large single bird.

(c) Small and medium flocking bird.

(d) Large flocking bird. An engine test will be performed as follows:

(1) Large flocking bird engine tests will be performed using the bird mass and weights in Table 4, and ingested at a bird speed of 200 knots.

(2) Prior to the ingestion, the engine must be stabilized at no less than the mechanical rotor speed of the first exposed stage or stages that, on a standard day, would produce 90 percent of the sea-level static maximum rated takeoff power or thrust.

(3) The bird must be targeted on the first exposed rotating stage or stages at a blade airfoil height of not less than 50 percent measured at the leading edge.

(4) Ingestion of a large flocking bird under the conditions prescribed in this paragraph must not cause any of the following:

(i) A sustained reduction of power or thrust to less than 50 percent of maximum rated takeoff power or thrust during the run-on segment specified under paragraph (d)(5)(i) of this section.

(ii) Engine shutdown during the required run-on demonstration specified in paragraph (d)(5) of this section.

(iii) The conditions specified in paragraph (b)(3) of this section.

(5) The following test schedule must be used:

(i) Ingestion followed by 1 minute without power lever movement.

(ii) Followed by 13 minutes at not less than 50 percent of maximum rated takeoff power or thrust.

(iii) Followed by 2 minutes between 30 and 35 percent of maximum rated takeoff power or thrust.

(iv) Followed by 1 minute with power or thrust increased from that set in paragraph (d)(5)(iii) of this section, by between 5 and 10 percent of maximum rated takeoff power or thrust.

(v) Followed by 2 minutes with power or thrust reduced from that set in paragraph (d)(5)(iv) of this section, by between 5 and 10 percent of maximum rated takeoff power or thrust.

(vi) Followed by a minimum of 1 minute at ground idle then engine shutdown. The durations specified are times at the defined conditions. Power lever movement between each condition will be 10 seconds or less, except that power lever movements allowed within paragraph (d)(5)(ii) of this section are not limited, and for setting power under paragraph (d)(5)(iii) of this section will be 30 seconds or less.

(6) Compliance with the large flocking bird ingestion requirements of this paragraph (d) may also be demonstrated by:

(i) Incorporating the requirements of paragraph (d)(4) and (d)(5) of this section, into the large single bird test demonstration specified in paragraph (b)(1) of this section; or

(ii) Use of an engine subassembly test at the ingestion conditions specified in paragraph (b)(1) of this section if:

(A) All components critical to complying with the requirements of paragraph (d) of this section are included in the subassembly test;

(B) The components of paragraph (d)(6)(ii)(A) of this section are installed in a representative engine for a run-on demonstration in accordance with paragraphs (d)(4) and (d)(5) of this section; except that section (d)(5)(i) is deleted and section (d)(5)(ii) must be 14 minutes in duration after the engine is started and stabilized; and

(C) The dynamic effects that would have been experienced during a full engine ingestion test can be shown to be negligible with respect to meeting the requirements of paragraphs (d)(4) and (d)(5) of this section.

(7) Applicants must show that an unsafe condition will not result if any engine operating limit is exceeded during the run-on period.

Table 4 to Sec. 33.76.—Large Flocking Bird Mass and Weight

Engine inlet throat area (square meters/square inches)	Bird quantity	Bird mass and weight (kg (lbs))
A < 2.50 (3875)	None	
2.50 (3875) ≤ A < 3.50 (5425)	1	1.85 (4.08)
3.50 (5425) ≤ A < 3.90 (6045)	1	2.10 (4.63)
3.90 (6045) ≤ A	1	2.50 (5.51)

Appendix 4:

NTSB Aircraft Bird Strike Recommendations from Accident Investigations

June 29, 1973: The NTSB recommends that the FAA disseminate by the widest possible distribution of FAA Advisory Circular 150/5200-3A dated March 2, 1972, Subject "Bird Hazards to Aircraft", to ensure that all airport operators, flying schools, fixed base operators, and airline transport, commercial and private pilots, and flight instructors are fully aware of the hazards to aviation associated with bird strikes.

July 8, 1996: The NTSB recommends that the FAA issue appropriate bulletins to urge pilots and maintenance personnel to report all bird strike incidents to the FAA Form 5200-7.

July 8, 1996: The NTSB recommends that the FAA develop and issue guidance to air traffic control terminal controllers to include specific info regarding the type, size and location of bird hazards on Automatic Terminal Information Service (ATIS).

July 8, 1996: The NTSB recommends that the FAA annually brief air traffic controllers on the importance of adhering to the guidance in paragraph 2-1-22 of FAA order 7110.65 "Air Traffic Control", regarding the dissemination of bird hazard info to pilots.

July 8, 1996: The NTSB recommends that the FAA develop a set of "scare tactic" procedures that can be requested by pilots, air traffic controllers, and/or airport personnel and executed by the proper personnel to disperse birds near runways. Disseminate these procedures to all parties in the appropriate manuals.

July 8, 1996: The NTSB recommends that the FAA revise the aeronautical info manual, paragraph 7-4-2, Section B, to advise pilots to delay takeoff whenever a bird hazard exists in the runway environment.

Nov 19, 1999: The NTSB recommends that the FAA evaluate the potential for using avian hazard advisory system technology for bird strike risk reduction in civil aviation and if found feasible, implement such a system in high-risk areas, such as major hub airports and along migratory bird routes nationwide.

Nov 19, 1999: The NTSB recommends that the FAA require all airplane operators to report bird strikes to the FAA.

Nov 19, 1999: The NTSB recommends that the FAA, before allowing high-speed, low-level airplane operations, evaluate the potential risk of increased bird strike hazards to air carrier turbojet airplanes.

Nov 19, 1999: The NTSB recommends that the FAA with representatives from the US Department of Agriculture, the Department of the Interior, the Department of Defense and the US Army Corps of Engineers, convene a task force to establish a permanent bird strike working group to facilitate conflict resolution and improve communication between aviation safety agencies and wildlife conservation interests.

Nov 19, 1999: The NTSB recommends to the US Department of Agriculture: participate in a task force, to be convened by the FAA, to establish a permanent bird strike working group to facilitate conflict resolution and improve communication between aviation safety agencies and wildlife conservation interests.

Sept 5, 2008: The NTSB recommends that the FAA evaluate the potential for using avian hazard technology for bird strike risk reduction in civil aviation and, if found feasible, implement such a system in high-risk areas, such as major hub airports and along migratory bird routes nationwide.

Sept 5, 2008: The NTSB recommends that the FAA in consultation with the US Department of Agriculture, require that wildlife assessments be conducted at all 14 Code of Federal Regulations Part 139 airports where such assessments have not already been conducted.

Sept 5, 2008: The NTSB recommends that the FAA ensure that the wildlife hazard management programs are incorporated into the airport certification manuals and periodically inspect the programs' progress.

Sept 5, 2008: The NTSB recommends that the FAA require all airplane operators to report bird strikes to the FAA.

Sept 5, 2008: The NTSB recommends that the FAA contract with an appropriate agency to provide proper identification of bird remains, establish timely procedures for proper bird species identification, and ensure that airport and aircraft maintenance employees are familiar with procedures.

Sept 5, 2008: The NTSB recommends that the FAA, before allowing high-speed, low-level airplane operations, evaluate the potential risk of increased bird strike hazards to air carrier turbojet airplanes.

Sept 5, 2008: The NTSB recommends that the FAA with representatives from the US Department of Agriculture, the

Department of the Interior, the Department of Defense and the US Army Corps of Engineers convene a task force to establish a permanent bird strike working group to facilitate conflict resolution and improve communication between aviation safety agencies and wildlife conservation interests.

Sept 5, 2008: The NTSB recommends to the US Department of Agriculture: participate in a task force, to be convened by the FAA, to establish a permanent bird strike working group to facilitate conflict resolution and improve communication between aviation safety agencies and wildlife conservation interests.

Sept 5, 2008: The NTSB recommends to the US Army Corps of Engineers: participate in a task force, to be convened by the FAA, to establish a permanent bird strike working group to facilitate conflict resolution and improve communication between aviation safety agencies and wildlife conservation interests.

Sept 5, 2008: The NTSB recommends to the US Department of Defense: participate in a task force, to be convened by the FAA, to establish a permanent bird strike working group to facilitate conflict resolution and improve communication between aviation safety agencies and wildlife conservation interests.

Sept 5, 2008: The NTSB recommends to the US Department of Interior: participate in a task force, to be convened by the FAA, to establish a permanent bird strike working group to facilitate conflict resolution and improve communication between aviation safety agencies and wildlife conservation interests.

July 28, 2009: The NTSB recommends that the bird strike certification requirements for 14 Code of Federal Regulations Part 25 airplanes be revised so that protection from in-flight impact with birds is consistent across all airframe structures. Consider the most current military and civilian bird strike database information and trends in bird populations in drafting this revision.

July 28, 2009: The NTSB recommends that the FAA verify that all federally obligated general aviation airports that are located near woodlands, water, wetlands, or other wildlife attractants are complying with the requirements to perform wildlife hazard assessments (WHA) as specified in Federal Aviation Administration Advisory Circular 150/5200-33B, Hazardous Wildlife Attractants on or Near Airports.

July 28, 2009: The NTSB recommends that the FAA require aircraft manufacturers to develop aircraft-specific guidance information that will assist pilots in devising precautionary aircraft operational strategies for minimizing the severity of aircraft damage sustained during a bird strike, should one occur, when operating in areas of known bird activity. This guidance information can include, but is not limited to, airspeed charts that depict minimum safe airspeeds for various aircraft gross weights, flap configurations, and power settings; and maximum airspeeds, defined as a function of bird masses, that are based on the aircraft's demonstrated bird strike energy.

July 28, 2009: The NTSB recommends that the FAA require all 14 Code of Federal Regulations (CFR) Part 139 airports and 14 CFR Part 121, Part 135, and Part 91 Subpart K aircraft operators to report all wildlife strikes, including species identification, if possible, to the Federal Aviation Administration National Wildlife Strike Database.

Appendix 5:

AIM Regulations for Ditching an Airliner & Ditching Demonstrations for Aircraft Certification

AIM Regulations for Ditching an Airliner
(from Airman's Information Manual, Section 6-3-2-c)

c. When in a distress condition with bailout, crash landing or ditching imminent, take the following additional actions to assist search and rescue units:

1. Time and circumstances permitting, transmit as many as necessary of the message elements in any of the following that you think might be helpful:

    (a) ELT status.
    (b) Visible landmarks.
    (c) Aircraft color.
    (d) Number of persons on board.
    (e) Emergency equipment on board.

2. Actuate your ELT if the installation permits.

3. For bailout, and for crash landing or ditching, if risk of fire is not a consideration, set your radio for continuous transmission.

4. If it becomes necessary to ditch, make every effort to ditch near a surface vessel. If time permits, an FAA facility should be able to get the position of the nearest commercial or Coast Guard vessel from a Coast Guard Rescue Coordination Center.

5. After a crash landing, unless you have good reason to believe that you will not be located by search aircraft or ground teams, it is best to remain with your aircraft and prepare means for signaling search aircraft.

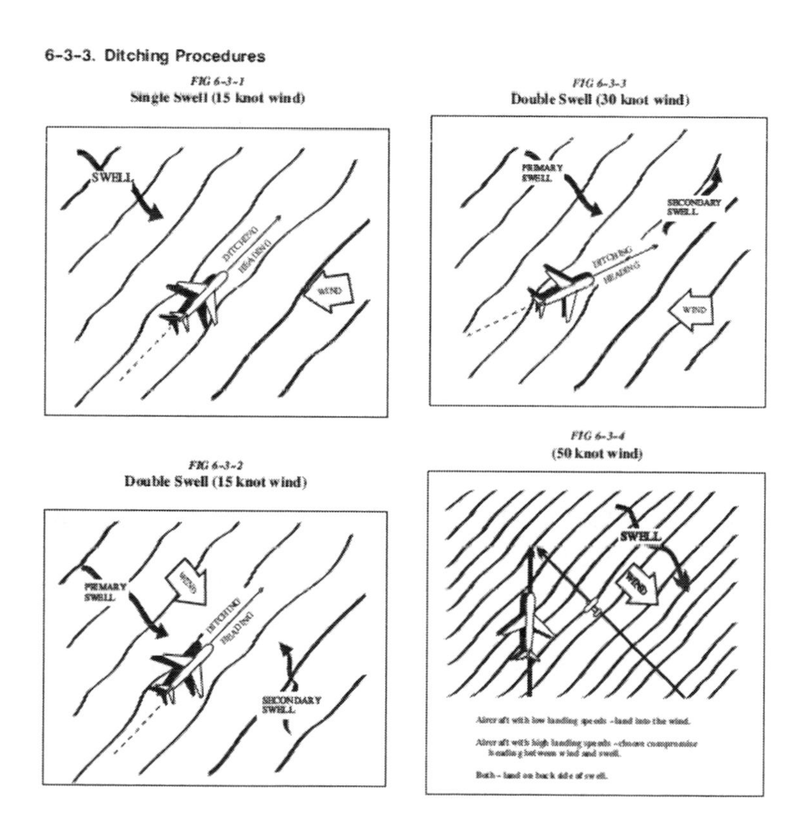

FIG 6-3-5
Wind-Swell-Ditch Heading

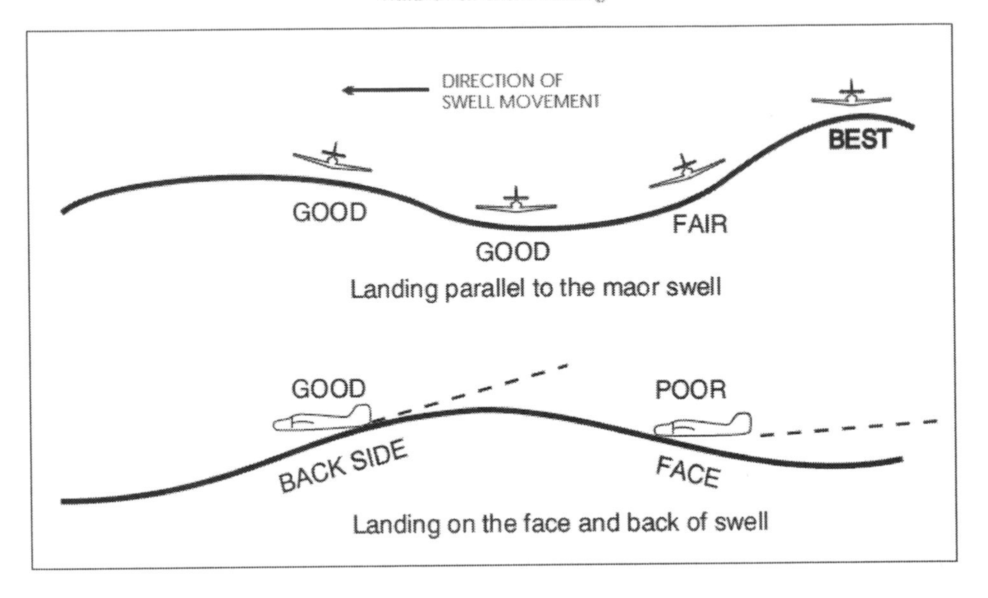

a. A successful aircraft ditching is dependent on three primary factors. In order of importance, they are:

1. Sea conditions and wind.

2. Type of aircraft.

3. Skill and technique of pilot.

b. Common oceanographic terminology.

1. Sea. The condition of the surface that is the result of both waves and swells.

2. Wave (or Chop). The condition of the surface caused by the local winds.

3. Swell. The condition of the surface which has been caused by a distance disturbance.

4. Swell Face. The side of the swell toward the observer. The backside is the side away from the observer. These

definitions apply regardless of the direction of swell movement.

5. Primary Swell. The swell system having the greatest height from trough to crest.

6. Secondary Swells. Those swell systems of less height than the primary swell.

7. Fetch. The distance the waves have been driven by a wind blowing in a constant direction, without obstruction.

8. Swell Period. The time interval between the passage of two successive crests at the same spot in the water, measured in seconds.

9. Swell Velocity. The speed and direction of the swell with relation to a fixed reference point, measured in knots. There is little movement of water in the horizontal direction. Swells move primarily in a vertical motion, similar to the motion observed when shaking out a carpet.

10. Swell Direction. The direction from which a swell is moving. This direction is not necessarily the result of the wind present at the scene. The swell may be moving into or across the local wind. Swells, once set in motion, tend to maintain their original direction for as long as they continue in deep water, regardless of changes in wind direction.

11. Swell Height. The height between crest and trough, measured in feet. The vast majority of ocean swells are lower than 12 to 15 feet, and swells over 25 feet are not common at any spot on the oceans. Successive swells may differ considerably in height.

c. In order to select a good heading when ditching an aircraft, a basic evaluation of the sea is required. Selection of a good ditching heading may well minimize damage and could save your life. It can be extremely dangerous to land into the wind without regard to sea conditions; the swell system, or systems, must be taken into consideration. Remember one axiom – AVOID THE FACE OF A SWELL.

1. In ditching parallel to the swell, it makes little difference whether touchdown is on the top of the crest or in the trough. It is preferable, however, to land on the top or back side of the swell, if possible. After determining which heading (and its reciprocal) will parallel the swell, select the heading with the most into-the-wind component.

2. If only one swell system exists, the problem is relatively simple – even with a high, fast system. Unfortunately, most cases involve two or more swell systems running in different directions. With more than one system present, the sea presents a confused appearance. One of the most difficult situations occurs when two swell systems are at right angles. For example, if one system is eight feet high, and the other three feet, plan to land parallel to the primary system, and on the down swell of the secondary system. If both systems are of equal height, a compromise may be advisable – select an intermediate heading at 45 degrees down swell to both systems. When landing down a secondary swell, attempt to touch down on the back side, not on the face of the swell.

3. If the swell system is formidable, it is considered advisable, in landplanes, to accept more crosswind in order to avoid landing directly into the swell.

4. The secondary swell system is often from the same direction as the wind. Here, the landing may be made parallel to the primary system, with the wind and secondary system at an angle. There is a choice to two directions paralleling the primary system. One direction is downwind and down the secondary swell, and the other is into the wind and into the secondary swell – the

choice will depend on the velocity of the wind versus the velocity and height of the secondary swell.

d. The simplest method of estimating the wind direction and velocity is to examine the wind streaks on the water. These appear as long streaks up- and downwind. Some persons may have difficulty determining wind direction after seeing the streaks on the water. Whitecaps fall forward with the wind but are overrun by the waves thus producing the illusion that the foam is sliding backward. Knowing this, and by observing the direction of the streaks, the wind direction is easily determined. Wind velocity can be estimated by noting the appearance of the whitecaps, foam and wind streaks.

1. The behavior of the aircraft on making contact with the water will vary within wide limits according to the state of the sea. If landed parallel to a single swell system, the behavior of the aircraft may approximate that to be expected on a smooth sea. If landed into a heavy swell or into a confused sea, the deceleration forces may be extremely great – resulting in breaking up of the aircraft. Within certain limits, the pilot is able to minimize these forces by proper sea evaluation and selection of ditching heading.

2. When on final approach, the pilot should look ahead and observe the surface of the sea. There may be shadows and whitecaps – signs of large seas. Shadows and whitecaps close together indicate short and rough seas. Touchdown in these areas is to be avoided. Select and touchdown in any area (only about 500 feet is needed) where the shadows and whitecaps are not so numerous.

3. Touchdown should be at the lowest speed and rate of descent which permit safe handling and optimum nose up attitude on impact. Once first impact has been made, there is often little the pilot can do to control a landplane.

e. Once pre-ditching preparations are completed, the pilot should turn to the ditching heading and commence let-down. The aircraft should be flown low over the water, and slowed down until ten knots or so above stall. At this point, additional power should be used to overcome the increased drag caused by the nose-up attitude. When a smooth stretch of water appears ahead, cut power, and touchdown at the best recommended speed as fully stalled as possible. By cutting power when approaching a relatively smooth area, the pilot will prevent overshooting and will touchdown with less chance of planing off into a second uncontrolled landing. Most experienced seaplane pilots prefer to make contact with the water in a semi-stalled attitude, cutting power as the tail makes contact. This technique eliminates the chance of misjudging altitude with a resultant heavy drop in a fully stalled condition. Care must be taken not to drop the aircraft from too high an altitude or to balloon due to excessive speed. The altitude above water depends on the aircraft. Over glassy smooth water, or at night without sufficient light, it is very easy for even the most experienced pilots to misjudge altitude by 50 feet or more. Under such conditions, carry enough power to maintain nine to twelve degrees nose-up attitude, and 10 to 20 percent over stalling speed until contact is made with the water. The proper use of power on the approach is of great importance. If power is available on one side only, a little power should be used to flatten the approach; however, the engine should not be used to such an extent that the aircraft cannot be turned against the good engines right down to the stall with a margin of rudder movement available. When near the stall, sudden application of excessive unbalanced power may result in loss of directional control. If power is available on one side only, a slightly higher than normal glide approach speed should be used. This will insure good control and some margin of speed after leveling off without excessive use of power. The use of power in ditching is so important that when it is certain that the coast cannot be reached, the pilot should, if possible, ditch before fuel is exhausted. The use of power in a night or instrument ditching is far more essential than under daylight contact conditions.

1. If no power is available, a greater than normal approach speed should be used down to the flare-out. This speed margin will allow the glide to be broken early and more gradually, thereby giving the pilot time and distance to feel for the surface – decreasing the possibility of stalling high or flying into the water. When landing parallel to a swell system, little difference is noted between landing on top of a crest or in the trough. If the wings of aircraft are trimmed to the surface of the sea rather than the horizon, there is little need to worry about a wing hitting a swell crest. The actual slope of a swell is very gradual. If forced to land into a swell, touchdown should be made just after passage of the crest. If contact is made on the face of the swell, the aircraft may be swamped or thrown violently into the air, dropping heavily into the next swell. If control surfaces remain intact, the pilot should attempt to maintain the proper nose-above-the-horizon attitude by rapid and positive use of the controls.

f. After Touchdown. In most cases, drift caused by crosswind can be ignored; the forces acting on the aircraft after touchdown are of such magnitude that drift will be only a secondary consideration. If the aircraft is under good control, the "crab" may be kicked out with rudder just prior to touchdown. This is more important with high-wing aircraft, for they are laterally unstable on the water in a crosswind and may roll to the side in ditching.

Ditching Demonstrations for Aircraft Certification
(from FAA Order 8099.1 Volume 3 Chapter 30 Section 4)
3-2531. General:

a. An applicant or certificate holder who proposes to operate a landplane in extended over-water operation must conduct a ditching demonstration. Extended over-water operation is defined as an operation over water at a horizontal distance of more than 50 NM from the nearest shoreline. However, in some cases, operators are allowed to operate certain types and models of airplanes at a distance greater than 50 miles from land without the operation being designated as an extended over-water operation. When this is the case, the waiver that allows this will be given in the operations specifications. The ditching demonstration is conducted in accordance with the requirements specified in Title 14 of the Code of Federal Regulations (14 CFR) part 121, §121.291(d) and (e), part 121, appendix D(b) or part 125, §125.189, part 125, appendix B, and the direction and guidance provided in this section. The purpose of the demonstration is to evaluate the operator's ability to safely prepare the passengers, airplane, and ditching equipment for a planned water landing. During the demonstration, the following four areas are evaluated:

Emergency training program
Ditching procedures
Crewmember competency
Equipment reliability and capability

b. Ditching and water landing are defined differently. Ditching, as commonly used in aviation, is a planned event. When the airplane lands in the water without warning, this is an unplanned water landing. A ditching demonstration will simulate a planned water landing. The preparation for ditching is similar in nature to the preparation for a planned evacuation.

3-2532. Crewmember Criteria: The selection and number of crewmembers to be used in the ditching demonstration is very important. The following paragraphs supplement the information in section 2, paragraph 3-2483F:

a. The qualifications of the crewmembers used in the ditching demonstration should be consistent with the qualification of line crewmembers. Whenever possible, crewmembers used in this demonstration should have been "line crewmembers" for the last 2 years. Experience gained prior to the previous 24 months should not be a consideration when selecting crewmembers for possible use in the demonstration. In addition, when possible, crewmembers should not have been used in a demonstration within the last 6 months. There are smaller airlines where this may not be possible. When this is the case, the former experience should be documented and included in the report regarding the demonstration. When the Federal Aviation Administration (FAA) determines that crewmembers to be used in the evacuation and ditching demonstrations have been allowed to "practice" opening the doors/exits, they should not allow these crewmembers to be used in the ditching demonstration, unless this additional training is included in the operator's FAA approved training program.

b. The air carrier should present a minimum of two complete crews for the demonstration(s). During the consultation with the Air Transportation Division (AFS-200) or the General Aviation and Commercial Division (AFS-800), through the CHDO that is required as a result of two demonstration failures, the lack of trained crewmembers for future demonstrations should be discussed. It is possible that, in the case of failure due to equipment, the same crewmembers can be used to test the equipment. However, this decision should be made in consultation with AFS-200 or AFS-800 as appropriate and the CHDO.

c. It is very important that the "back-up" crewmembers that may be used if the first demonstration fails are not given any information about the first demonstration. Sometimes, this is best accomplished by having these crewmembers isolated in an area which is physically removed from the first ditching demonstration. However, if these back-up crewmembers are not held in an area away from the demonstration, they should stay in a group with an FAA inspector present so the inspector can insure they are not given any information about the first demonstration.

d. When an airline is new, typical line crewmembers may not be available. When this is the case, the carrier must train the first cadre of flight attendants; it is quite possible that these flight attendants will also be instructors. Nevertheless, they should not be given instruction or experience that will not be given to trainees who will be expected to serve as flight attendants on this aircraft in operations. For example, they should not have had "train the trainer" training until after their participation in the ditching demonstration. Flight attendant managers who are in charge of the air carrier's overall flight attendant program should not be used as crewmembers during the demonstration, unless no other flight attendants have been hired.

e. Crewmembers who are used in evacuation demonstrations may also be used in the ditching demonstration. However, the FAA encourages, whenever possible the use of separate crewmembers for the emergency evacuation demonstration and the ditching demonstration. Some air carriers may not be aware of the stress level the crews face by participating in these types of demonstrations. Additionally, by providing separate crewmembers for each demonstration, it provides the FAA with a better assessment of the training program. In the event of a non-flight attendant demonstration failure (e.g., equipment failure), it is recommended that a new flight attendant crew be selected from the remaining flight attendant group.

3-2533. Regulatory Requirements: Section 121.291(d) or §125.189(c) requires an operator to conduct a ditching demonstration for each type and model of airplane used in extended overwater operations. This demonstration must be conducted in accordance with each of the conditions stipulated in §121.291(d) or §125.189(c) as appropriate unless the operator can present documentation that another operator conducting operations using part 121 operations specifications has conducted a successful ditching demonstration using the same type and model of airplane. When the operator provides the FAA with this documentation, then §121.291(e) provides relief from some of the conditions contained in §121.291(d) . Specifically, §121.291(e) provides relief from those requirements of paragraphs (b)(2), (b)(4) and (b)(5) of appendix D to part 121. The purpose of a ditching demonstration is to show the FAA that the certificate holder has the ability to efficiently carry out its ditching procedures. In accordance with §121.291 and appendix D to part 121 or §125.189 and appendix B to part 125, the ditching demonstration is conducted as follows:

a. The demonstration must be conducted during daylight hours or in a lighted hangar if conducted at night.

b. The FAA minimum required crew complement (both flight crew and flight attendants) must be available and used during the ditching demonstration.

c. When an operator's procedures use able-bodied persons (ABPs) to remove or launch life rafts, then the same number of persons who will act as ABP(s) must be used in the demonstration. The ABP(s) should be provided by the operator and should have experiences similar to average passengers. Crewmembers, mechanics, and other air carrier personnel who would have knowledge regarding the use of emergency equipment should not be used. The FAA should insure that ABP(s) are not given additional training. The ABP(s) should be briefed and perform the duties as stipulated in the appropriate crewmember manuals. The operator should supply enough ABP(s) to ensure coverage if the first demonstration fails.

d. Stands must be placed at each emergency exit and wing. The life rafts or slide rafts should be inflated onto the stands and then lowered to the hangar floor. This will prevent injury to participants, as well as damage to the life rafts or slide rafts.

NOTE: 14 CFR part 25, §25.807(d) requires that, during type certification, ditching emergency exits must be above the calculated waterline which will exist when the airplane is at rest in the water. This "waterline" and the designated ditching emergency exits are defined in the "manufacturer's ditching document", which is part of the final part 25 type certification report. The operator should obtain waterline and ditching exit information from the manufacturer. This waterline is where the tops of the stands should be positioned.

e. Each evacuee (crewmembers and passenger participants (ABPs), if applicable) must don and inflate a life preserver according to the operator's procedures and the flight attendant's briefing.

f. Each life raft or slide raft must be launched and inflated according to the operator's procedures. When air carrier procedures require a survival kit to be attached to a door-mounted slide/raft, the survival kit must be attached prior to inflation. All required emergency equipment must be placed in the rafts. Each evacuee must enter a life raft or slide raft. The crewmembers assigned to the raft shall locate and describe the use of each piece of emergency equipment.

g. Each life raft must be removed from stowage for inspection. One life raft or slide raft (selected by the FAA) shall be inflated and launched, and the evacuees assigned to that raft shall get in it. The rafts and/or slide rafts must be the same as those used on the aircraft.

Note: In addition, they must be equipped as they would be on the aircraft in regular operations. The crewmembers assigned to the raft shall locate and describe the use of each item of emergency equipment.

h. Either the airplane, a life-size mockup, or a floating device that accurately simulates the passenger compartment must be used for the demonstration (part 121, appendix D(b)(6)(i)(ii) or part 125, appendix B(b)(6)(i)(ii)).

Note: It is FAA policy to use an airplane for all ditching demonstrations. If the operator proposes to use a life-size mockup or a floating device to conduct the demonstration, approval must be granted by AFS-200 or AFS-800 as appropriate.

3-2534. The Ditching Demonstration Plan: The ditching demonstration is normally conducted after the satisfactory completion of the aborted takeoff emergency evacuation demonstrations. In these situations, the same team leader and FAA team members should conduct and observe the ditching demonstration. Every effort should be made to have cabin safety inspectors (CSI) act as team leaders or at least be involved in the planning and conducting of the ditching demonstration. This may not always be possible because of the workload of some cabin safety inspectors. When a cabin safety inspector is not available, assistance from other experienced inspectors should be sought. However, if an operator plans to initiate flights into extended-overwater areas for the first time with an airplane that they previously operated over land areas, they must conduct a ditching demonstration. See paragraph 3-2533.

a. If the operator plans to conduct the ditching demonstration in conjunction with the emergency evacuation aborted takeoff demonstration, the operator's aborted takeoff demonstration plan must include information applicable to the ditching demonstration such as the following:

1) Copies of the operator's manual relating to crewmember's ditching duties and responsibilities.

2) A description of applicable emergency equipment used for ditching (such as life rafts, survival gear) including the type and model of the emergency equipment.

b. If the operator presents the FAA with documentation that another operator conducting operations using part 121 operations specifications has conducted a successful ditching demonstration using the same type and model of airplane, then §121.291(e) provides relief from some of the conditions contained in §121.291(d). Specifically, § 121.291(e) provides relief from those requirements of paragraphs (b)(2), (b)(4) and (b)(5) of appendix D to part 121.

c. If the operator must conduct a ditching demonstration that is not in conjunction with an emergency evacuation aborted takeoff demonstration, the operator's demonstration plan must be submitted at least 15 working days before the date of the actual demonstration. This plan must include the following information:

1) The airplane type and model which will be used;

2) The proposed date, time, and location of the ditching demonstration;

3) The name and telephone number of the company's ditching demonstration coordinator;

4) A representative diagram of the aircraft which includes the following:

a) Location and designation of each exit
b) Location of each item of emergency ditching equipment including:
Life rafts/slide rafts
Survival radios
Pyrotechnic signaling devices
Passenger/crewmember life preservers or individual floatation devices

5) A list of all crewmembers (both flight crew and flight attendants) that are qualified to participate in the demonstration must be in the operator's plan. The crewmembers must be qualified in the aircraft to be used; however, the initial operating experience requirement need not be completed. Flight attendant personnel (in accordance with §121.291(c)(3) must have completed an FAA-approved training program for the type and model of airplane being demonstrated. For part 125, the training required by that part must be completed. Flight attendants designated by the FAA to participate in the demonstration shall not be provided emergency training or aircraft emergency equipment familiarization more than specified in the operator's approved training program before the demonstration.

6) Copies of the appropriate crewmember manual pages

describing ditching duties and responsibilities, including cabin preparation time parameters for both planned and unplanned ditching. See subparagraph 3-2533A for guidance regarding ABPs.

3-2535. Review of the Ditching Demonstration Plan:

a. When the ditching demonstration plan has been submitted, the principal operations inspector and cabin safety inspector must review the proposal to ensure the following:

1) The proposed demonstration will meet the criteria in §121.291(d) or (e) and part 121, appendix D(b). For part 125, the proposed demonstration will meet the criteria in §125.189(c) and part 125, appendix B(b) or in accordance with the deviation authority granted under §125.3. (See Related Task #73 and Notice 8700.46, Evaluate an Application for Deviation or Special Authorization for part 125.)

2) The emergency training program and ditching procedures in the operator's manual must have been approved and accepted and provide for safe operating practices.

3) The ditching duties and responsibilities, including cabin preparation time parameters for both planned and unplanned ditching is realistic and is understood by all.

b. The FAA team must plan for the observation and evaluation of the ditching demonstration. Normally, the demonstration is conducted after the completion of a successful aborted takeoff emergency evacuation demonstration. If an aborted takeoff emergency evacuation demonstration is not conducted, the district office manager shall appoint an FAA ditching demonstration team and a team leader in the same manner as the aborted takeoff demonstration. As a reminder, every effort should be made to have CSIs act as team leaders or at least be involved in the planning and conducting of the ditching demonstration.

3-2536. Conduct of the Ditching Demonstration: The ditching demonstration shall be conducted in the following manner:

a. Before the ditching demonstration, the team shall inspect each item of emergency ditching equipment for compliance with appropriate airworthiness and other relevant directives.

b. The FAA team leader ensures inspectors and crewmembers are at their assigned positions and then advises the captain to commence the demonstration.

c. The amount of time a crew is given to prepare the cabin for a ditching demonstration should be reasonable. Certificate holder manuals and procedures stipulate crewmember notification, including time parameters for both planned and unplanned ditching, before touchdown, etc., in their crewmember operating manuals. Both the FAA team leader and the operator should agree on a time limitation for the demonstration based on the operator's planned ditching time parameters. The flight attendants should be performing duties associated with normal flight, such as serving meals with the cart in the aisle when the signal to ditch is given. The timing should start when the pilot notifies the flight attendants of the impending ditch. The crewmembers must use the air carrier's procedures as outlined in the appropriate manuals. The timing stops when the flight attendants have completed preparations and notify the captain that the cabin is prepared. It is important that inspectors evaluating ditching procedures ensure that both timing and preparations follow those stipulated in the operator's manuals.

Note: Regulations do not specify a maximum time limit for the demonstration. However, it is imperative that emergency equipment, crewmember competency, and emergency procedures provide for rapid evacuation since during an actual ditching situation, the airplane may remain afloat for only a short time. During the demonstration, emphasis is on crewmember ability and efficiency in the time period between the decision to ditch and the actual water landing. Fifteen minutes is considered a realistic time acceptable for ditching preparation beginning with the ditching announcement to the simulated water landing. However, timing adjustments may be made in coordination with the FAA team leader and operator, as some operator's manuals stipulate longer or shorter time periods for planned preparations. Once the time is agreed upon, all participating crewmembers must correctly don life preservers, brief passenger participants (ABPs) (if applicable), secure the cabin, and complete all required checklists and procedures within the time specified. Failure to be prepared at the end of this time constitutes an unsatisfactory demonstration.

d. The FAA team leader begins timing when the captain issues the prepare-for-ditching order. At the end of the fifteen minutes, or other agreed upon time for the planned ditching demonstration, the crew must be prepared for a simulated water landing. After the simulated aircraft landing and stopping, each crewmember must follow the operator's procedures as contained in the appropriate manuals regarding the launching and boarding of life rafts and/or slide rafts. If the aircraft has more than one type of raft, then each type should be launched, but only one type needs to be boarded. After the timing has stopped, each crewmember must board the raft and must be able to answer questions regarding the location and function of various pieces of equipment on the raft, describe the use of each item in the survival kit, and erect the canopy as a group. The current regulations require that training programs ensure that each crewmember remains adequately trained and currently proficient with respect to each airplane, crewmember position, and type of operation in which he or she serves. If a crewmember fails to answer the appropriate questions, then the team leader should carefully evaluate the question and failure to answer so that appropriate changes to the operator's training program can be made, if needed. However, the failure of one crewmember to answer a question may not constitute failure of the demonstration.

e. Section 121.291(d) requires that all life rafts and slide rafts be launched and inflated. Section 121.291(e) requires only one life raft (or slide raft), designated by the FAA team leader, to be launched and inflated. However, if a slide raft is the primary means of flotation, then this should be the selected raft. For the purpose of this demonstration, "launching" a life raft means to remove it from stowage, manipulate it out of the airplane (via stands or ramps), and position it on the ground before inflation. "Launching" a slide raft means to inflate it in a normal manner and then lower it to the ground.

Note: Section 121.291(e) does not require detachment of each slide raft from its respective door mounting. However, each slide raft must be inspected for its airworthiness. Any life rafts stowed inside the airplane must be removed from stowage and placed on the cabin floor for inspection.

f. When an operator's procedures use ABPs to remove or launch life rafts, then the same number of ABPs should be used in the demonstration. The ABPs should be provided by the operator and should have experiences similar to average passengers. Crewmembers, mechanics, and other air carrier personnel who would have knowledge regarding the use of emergency equipment should not be used. The FAA should insure that ABPs are not given additional training. The ABPs should be briefed and perform the duties as stipulated in the appropriate crewmember manuals. The operator should supply enough ABPs to ensure coverage if the first demonstration fails.

Appendix 6:

Commercial Airport Certification (FAR Part 139)

FAR Part 139 Subpart B Certification
Sec. 139.101 - Certification requirements: General.

(a) No person may operate a land airport in any State of the United States, the District of Columbia, or any territory or possession of the United States, serving any scheduled passenger operation of an air carrier operating an aircraft having a seating capacity of more than 30 passengers without an airport operating certificate, or in violation of that certificate, the applicable provisions of this part, or the approved airport certification manual for that airport.

(b) Unless otherwise authorized by the Administrator, no person may operate a land airport in any State of the United States, the District of Columbia, or any territory or possession of the United States, serving any unscheduled passenger operation of an air carrier operating an aircraft having a seating capacity of more than 30 passengers without a limited airport operating certificate, or in violation of that certificate, the applicable provisions of this part, or the approved airport specifications for that airport.

FAR Part 139 Subpart D Operations
Sec. 139.305 - Paved areas.

(a) Each certificate holder shall maintain, and promptly repair the pavement of, each runway, taxiway, loading ramp, and parking area on the airport which is available for air carrier use as follows:

(1) The pavement edges shall not exceed 3 inches difference in elevation between abutting pavement sections and between full strength pavement and abutting shoulders.

(2) The pavement shall have no hole exceeding 3 inches in depth nor any hole the slope of which from any point in the hole to the nearest point at the lip of the hole is 45 degrees or greater as measured from the pavement surface plane, unless, in either case, the entire area of the hole can be covered by a 5-inch diameter circle.

(3) The pavement shall be free of cracks and surface variations which could impair directional control of air carrier aircraft.

(4) Except as provided in paragraph (b) of this section, mud, dirt, sand, loose aggregate, debris, foreign objects, rubber deposits, and other contaminants shall be removed promptly and as completely as practicable.

(5) Except as provided in paragraph (b) of this section, any chemical solvent that is used to clean any pavement area shall be removed as soon as possible, consistent with the instructions of the manufacturer of the solvent.

(6) The pavement shall be sufficiently drained and free of

depressions to prevent ponding that obscures markings or impairs safe aircraft operations.

(b) Paragraphs (a)(4) and (a)(5) of this section do not apply to snow and ice accumulations and their control, including the associated use of materials such as sand and deicing solutions.

(c) FAA Advisory Circulars in the 150 series contain standards and procedures for the maintenance and configuration of paved areas which are acceptable to the Administrator.

Sec. 139.307 - Unpaved areas.

(a) Each certificate holder shall maintain and promptly repair the surface of each gravel, turf, or other unpaved runway, taxiway, or loading ramp and parking area on the airport which is available for air carrier use as follows:

(1) No slope from the edge of the full-strength surfaces downward to the existing terrain shall be steeper than 2:1.

(2) The full-strength surfaces shall have adequate crown or grade to assure sufficient drainage to prevent ponding.

(3) The full-strength surfaces shall be adequately compacted and sufficiently stable to prevent rutting by aircraft, or the loosening or buildup of surface material which could impair directional control of aircraft or drainage.

(4) The full-strength surfaces must have no holes or depressions which exceed 3 inches in depth and are of a breadth capable of impairing directional control or causing damage to an aircraft.

(5) Debris and foreign objects shall be promptly removed from the surface.

(b) Standards and procedures for the maintenance and configuration of unpaved full-strength surfaces shall be included in the airport certification manual or the airport certification specifications, as appropriate, for compliance with this section.

Sec. 139.309 - Safety areas.

(a) To the extent practicable, each certificate holder shall provide and maintain for each runway and taxiway which is available for air carrier use --

(1) If the runway or taxiway had a safety area on December 31, 1987, and if no reconstruction or significant expansion of the runway or taxiway was begun on or after January 1, 1988, a safety area of at least the dimensions that existed on December 31, 1987; or

(2) If construction, reconstruction, or significant expansion of the runway or taxiway began on or after January 1, 1988, a safety area which conforms to the dimensions acceptable to the Administrator at the time construction, reconstruction, or expansion began.

(b) Each certificate holder shall maintain its safety areas as follows:

(1) Each safety area shall be cleared and graded, and have no potentially hazardous ruts, humps, depressions, or other surface variations.

(2) Each safety area shall be drained by grading or storm sewers to prevent water accumulation.

(3) Each safety area shall be capable under dry conditions of supporting snow removal equipment, and aircraft rescue and firefighting equipment, and supporting the occasional passage of aircraft without causing major damage to the aircraft.

(4) No object may be located in any safety area, except for objects that need to be located in a safety area because of their function. These objects shall be constructed, to the extent practical, on frangibly mounted structures of the lowest practical height with the frangible point no higher than 3 inches above grade.

(c) FAA Advisory Circulars in the 150 series contain standards and procedures for the configuration and maintenance of safety areas acceptable to the Administrator.

Sec. 139.311 - Marking and lighting.

(a) Each certificate holder shall provide and maintain at least the following marking systems for air carrier operations on the airport:

(1) Runway markings meeting the specifications for the approach with the lowest minimums authorized for each runway.

(2) Taxiway centerline and edge markings.

(3) Signs identifying taxiing routes on the movement area.

(4) Runway holding position markings and signs.

(5) ILS critical area markings and signs.

(b) Each certificate holder shall provide and maintain, when the airport is open during hours of darkness or during conditions below VFR minimums, at least the following lighting systems for air carrier operations on the airport:

(1) Runway lighting meeting the specifications for the approach with the lowest minimums authorized for each runway.

(2) One of the following taxiway lighting systems:

(i) Centerline lights.

(ii) Centerline reflectors.

(iii) Edge lights.

(iv) Edge reflectors.

(3) An airport beacon.

(4) Approach lighting meeting the specifications for the approach with the lowest minimums authorized for each runway, unless otherwise provided and maintained by the FAA or another agency.

(5) Obstruction marking and lighting, as appropriate, on each object within its authority which constitutes an obstruction under part 77 of this chapter. However, this lighting and marking is not required if it is determined to be unnecessary by an FAA aeronautical study.

(c) Each certificate holder shall properly maintain each marking or lighting system installed on the airport which is owned by the certificate holder. As used in this section, to "properly maintain" includes: To clean, replace, or repair any faded, missing, or nonfunctional item of lighting; to keep each item unobscured and clearly visible; and to ensure that each item provides an accurate reference to the user.

(d) Each certificate holder shall ensure that all lighting on the airport, including that for aprons, vehicle parking areas, roadways, fuel storage areas, and buildings, is adequately adjusted or shielded to prevent interference with air traffic control and aircraft operations.

(e) FAA Advisory Circulars in the 150 series contain standards and procedures for equipment, material, installation, and maintenance of light systems and marking listed in this section which is acceptable to the Administrator.

(f) Notwithstanding paragraph (a) of this section, a certificate holder is not required to provide the identified signs in paragraph (a)(3) of this section until January 1, 1995. Each certificate holder shall maintain each marking system that meets paragraph (a)(3) of this section.

Sec. 139.313 - Snow and ice control.

(a) Each certificate holder whose airport is located where snow and icing conditions regularly occur shall prepare, maintain, and carry out a snow and ice control plan.

(b) The snow and ice control plan required by this section shall include instructions and procedures for --

(1) Prompt removal or control, as completely as practical, of snow, ice, and slush on each movement area;

(2) Positioning snow off of movement area surfaces so that all air crarrier aircraft propellers, engine pods, rotors, and wingtips will clear any snowdrift and snowbank as the aircraft's landing gear traverses any full strength portion of the movement area;

(3) Selection and application of approved materials for snow and ice control to ensure that they adhere to snow and ice sufficiently to minimize engine ingestion;

(4) Timely commencement of snow and ice control operations; and

(5) Prompt notification, in accordance with §139.339, of all air carriers using the airport when any portion of the movement area normally available to them is less than satisfactorily cleared for safe operation by their aircraft.

(c) FAA Advisory Circulars in the 150 series contain standards for snow and ice control equipment, materials, and procedures for snow and ice control which are acceptable to the Administrator.

Sec. 139.319 - Aircraft rescue and firefighting: Operational requirements.

(a) Except as provided in paragraph (c) of this section, each certificate holder shall provide on the airport, during air carrier operations at the airport, at least the rescue and firefighting capability specified for the Index required by §139.317.

(b) *Increase in Index.* Except as provided in paragraph (c) of this section, if an increase in the average daily departures or the length of air carrier aircraft results in an increase in the Index required by paragraph (a) of this section, the certificate holder shall comply with the increased requirements.

(c) *Reduction in rescue and firefighting.* During air carrier operations with only aircraft shorter than the Index aircraft group required by paragraph (a) of this section, the certificate holder may reduce the rescue and firefighting to a lower level corresponding to the Index group of the longest air carrier aircraft being operated.

(d) Any reduction in the rescue and firefighting capability from the Index required by paragraph (a) of this section in accordance with paragraph (c) of this section shall be subject to the following conditions:

(1) Procedures for, and the persons having the authority to implement, the reductions must be included in the airport certification manual.

(2) A system and procedures for recall of the full aircraft rescue and firefighting capability must be included in the airport certification manual.

(3) The reductions may not be implemented unless notification to air carriers is provided in the Airport/Facility Directory or Notices to Airmen (NOTAM), as appropriate, and by direct notification of local air carriers.

(e) *Vehicle communications.* Each vehicle required under §139.317 shall be equipped with two-way voice radio communications which provides for contact with at least --

(1) Each other required emergency vehicle;

(2) The air traffic control tower, if it is located on the airport; and

(3) Other stations, as specified in the airport emergency plan.

(f) *Vehicle marking and lighting.* Each vehicle required under §139.317 shall --

(1) Have a flashing or rotating beacon; and

(2) Be painted or marked in colors to enhance contrast with the background environment and optimize daytime and nighttime visibility and identification.

(g) FAA Advisory Circulars in the 150 series contain standards for painting, marking and lighting vehicles used on airports which are acceptable to the Administrator.

(h) *Vehicle readiness.* Each vehicle required under §139.317 shall be maintained as follows:

(1) The vehicle and its systems shall be maintained so as to be operationally capable of performing the functions required by this subpart during all air carrier operations.

(2) If the airport is located in a geographical area subject to prolonged temperatures below 33 degrees Fahrenheit, the vehicles shall be provided with cover or other means to ensure equipment operation and discharge under freezing conditions.

(3) Any required vehicle which becomes inoperative to the extent that it cannot perform as required by §139.319(h)(1) shall be replaced immediately with equipment having at least equal capabilities. If replacement equipment is

not available immediately, the certificate holder shall so notify the Regional Airports Division Manager and each air carrier using the airport in accordance with §139.339. If the required Index level of capability is not restored within 48 hours, the airport operator, unless otherwise authorized by the Administrator, shall limit air carrier operations on the airport to those compatible with the Index corresponding to the remaining operative rescue and firefighting equipment.

(i) *Response requirements.* (1) Each certificate holder, with the airport rescue and firefighting equipment required under this part and the number of trained personnel which will assure an effective operation, shall --

(i) Respond to each emergency during periods of air carrier operations; and

(ii) When requested by the Administrator, demonstrate compliance with the response requirements specified in this section.

(2) The response required by paragraph (i)(1)(ii) of this section shall achieve the following performance:

(i) Within 3 minutes from the time of the alarm, at least one required airport rescue and firefighting vehicle shall reach the midpoint of the farthest runway serving air carrier aircraft from its assigned post, or reach any other specified point of comparable distance on the movement area which is available to air carriers, and begin application of foam, dry chemical, or halon 1211.

(ii) Within 4 minutes from the time of alarm, all other required vehicles shall reach the point specified in paragraph (i)(2)(i) of this section from their assigned post and begin application of foam, dry chemical, or halon 1211.

(j) *Personnel.* Each certificate holder shall ensure the following:

(1) All rescue and firefighting personnel are equipped in a manner acceptable to the Administrator with protective clothing and equipment needed to perform their duties.

(2) All rescue and firefighting personnel are properly trained to perform their duties in a manner acceptable to the Administrator. The training curriculum shall include initial and recurrent instruction in at least the following areas:

(i) Airport familiarization.

(ii) Aircraft familiarization.

(iii) Rescue and firefighting personnel safety.

(iv) Emergency communications systems on the airport, including fire alarms.

(v) Use of the fire hoses, nozzles, turrets, and other appliances required for compliance with this part.

(vi) Application of the types of extinguishing agents required for compliance with this part.

(vii) Emergency aircraft evacuation assistance.

(viii) Firefighting operations.

(ix) Adapting and using structural rescue and firefighting equipment for aircraft rescue and firefighting.

(x) Aircraft cargo hazards.

(xi) Familiarization with firefighters' duties under the airport emergency plan.

(3) All rescue and firefighting personnel participate in at least one live-fire drill every 12 months.

(4) After January 1, 1989, at least one of the required personnel on duty during air carrier operations has been trained and is current in basic emergency medical care. This training shall include 40 hours covering at least the following areas:

(i) Bleeding.

(ii) Cardiopulmonary resuscitation.

(iii) Shock.

(iv) Primary patient survey.

(v) Injuries to the skull, spine, chest, and extremities.

(vi) Internal injuries.

(vii) Moving patients.

(viii) Burns.

(ix) Triage.

(5) Sufficient rescue and firefighting personnel are available during all air carrier operations to operate the vehicles, meet the response times, and meet the miminum agent discharge rates required by this part;

(6) Procedures and equipment are established and maintained for alerting rescue and firefighting personnel by siren, alarm, or other means acceptable to the Administrator, to any existing or impending emergency requiring their assistance.

(k) *Emergency access roads.* Each certificate holder shall ensure that roads which are designated for use as emergency access roads for aircraft rescue and firefighting vehicles are maintained in a condition that will support those vehicles during all-weather conditions.

Sec. 139.321 - Handling and storing of hazardous substances and materials.

(a) Each certificate holder which acts as a cargo handling agent shall establish and maintain procedures for the protection of persons and property on the airport during the handling and storing of any material regulated by the Hazardous Materials Regulations (49 CFR part 171, *et seq.*), that is, or is intended to be, transported by air. These procedures shall provide for at least the following:

(1) Designated personnel to receive and handle hazardous substances and materials.

(2) Assurance from the shipper that the cargo can be handled safely, including any special handling procedures required for safety.

(3) Special areas for storage of hazardous materials while on the airport.

(b) Each certificate holder shall establish and maintain standards acceptable to the Administrator for protecting against fire and explosions in storing, dispensing, and otherwise handling fuel, lubricants, and oxygen (other than articles and materials that are, or are intended to be, aircraft cargo) on the airport. These standards shall cover facilities, procedures, and personnel training and shall address at least the following:

(1) Grounding and bonding.

(2) Public protection.

(3) Control of access to storage areas.

(4) Fire safety in fuel farm and storage areas.

(5) Fire safety in mobile fuelers, fueling pits, and fueling cabinets.

(6) After January 1, 1989, training of fueling personnel in fire safety in accordance with paragraph (e) of this section.

(7) The fire code of the public body having jurisdiction over the airport.

(c) Each certificate holder shall, as a fueling agent, comply with and, except as provided in paragraph (h) of this section, require all other fueling agents operating on the airport to comply with the standards established under paragraph (b) of this section and shall perform reasonable surveillance of all fueling activities on the airport with respect to those standards.

(d) Each certificate holder shall inspect the physical facilities of each airport tenant fueling agent at least once every 3 months for compliance with paragraph (b) of this section and maintain a record of that inspection for at least 12 months. The certificate holder may use an independent organization to perform this inspection if --

(1) It is acceptable by the Administrator; and

(2) It prepares a record of its inspection sufficiently detailed to assure the certificate holder and the FAA that the inspection is adequate.

(e) The training required in paragraph (b)(6) of this section shall include at least the following:

(1) At least one supervisor with each fueling agent shall have completed an aviation fuel training course in fire safety which is acceptable to the Administrator.

(2) All other employees who fuel aircraft, accept fuel shipments, or otherwise handle fuel shall receive at least on-the-job training in fire safety from the supervisor trained in accordance with paragraph (e)(1) of this section.

(f) Each certificate holder shall obtain certification once a year from each airport tenant fueling agent that the training required by paragraph (e) of this section has been accomplished.

(g) Unless otherwise authorized by the Administrator, each certificate holder shall require each tenant fueling agent to take immediate corrective action whenever the certificate holder becomes aware of noncompliance with a standard required by paragraph (b) of this section. The certificate holder shall notify the appropriate FAA Regional Airports Division Manager immediately when noncompliance is discovered and corrective action cannot be accomplished within a reasonable period of time.

(h) A certificate holder need not require an air carrier operating under part 121 or part 135 of this chapter to comply with the standards required by this section.

(i) FAA Advisory Circulars in the 150 Series contain standards and procedures for the handling and storage of hazardous substances and materials which are acceptable to the Administrator.

Sec. 139.323 - Traffic and wind direction indicators.

Each certificate holder shall provide the following on its airport:

(a) A wind cone that provides surface wind direction information visually to pilots. For each airport in a Class B airspace area, supplemental wind cones must be installed at each runway end or at least at one point visible to the pilot while on final approach and prior to takeoff. If the airport is open for air carrier operations during hours of darkness, the wind direction indicators must be lighted.

(b) For airports serving any air carrier operation when there is no control tower operating, a segmented circle around one wind cone and a landing strip and traffic pattern indicator for each runway with a right-hand traffic pattern.

Sec. 139.325 - Airport emergency plan.

(a) Each certificate holder shall develop and maintain an airport emergency plan designed to minimize the possibility and extent of personal injury and property damage on the airport in an emergency. The plan must include --

(1) Procedures for prompt response to all of the emergencies listed in paragraph (b) of this section, including a communications network; and

(2) Sufficient detail to provide adequate guidance to each person who must implement it.

(b) The plan required by this section must contain instructions for response to --

(1) Aircraft incidents and accidents;

(2) Bomb incidents, including designated parking areas for the aircraft involved;

(3) Structural fires;

(4) Natural disaster;

(5) Radiological incidents;

(6) Sabotage, hijack incidents, and other unlawful interference with operations;

(7) Failure of power for movement area lighting; and

(8) Water rescue situations.

(c) The plan required by this section must address or include --

(1) To the extent practicable, provisions for medical

services including transportation and medical assistance for the maximum number of persons that can be carried on the largest air carrier aircraft that the airport reasonably can be expected to serve;

(2) The name, location, telephone number, and emergency capability of each hospital and other medical facility, and the business address and telephone number of medical personnel on the airport or in the communities it serves, agreeing to provide medical assistance or transportation;

(3) The name, location, and telephone number of each rescue squad, ambulance service, military installation, and government agency on the airport or in the communities it serves, that agrees to provide medical assistance or transportation;

(4) An inventory of surface vehicles and aircraft that the facilities, agencies, and personnel included in the plan under paragraphs (c)(2) and (c)(3) of this section will provide to transport injured and deceased persons to locations on the airport and in the communities it serves;

(5) Each hangar or other building on the airport or in the communities it serves that will be used to accommodate uninjured, injured, and deceased persons;

(6) Crowd control, specifying the name and location of each safety or security agency that agrees to provide assistance for the control of crowds in the event of an emergency on the airport; and

(7) The removal of disabled aircraft including to the extent practical the name, location and telephone numbers of agencies with aircraft removal responsibilities or capabilities.

(d) The plan required by this section must provide for --

(1) The marshalling, transportation, and care of ambulatory injured and uninjured accident survivors;

(2) The removal of disabled aircraft;

(3) Emergency alarm systems; and

(4) Coordination of airport and control tower functions relating to emergency actions.

(e) The plan required by this section shall contain procedures for notifying the facilities, agencies, and personnel who have responsibilities under the plan of the location of an aircraft accident, the number of persons involved in that accident, or any other information necessary to carry out their responsibilities, as soon as that information is available.

(f) The plan required by this section shall contain provisions, to the extent practicable, for the rescue of aircraft accident victims from significant bodies of water or marsh lands adjacent to the airport which are crossed by the approach and departure flight paths of air carriers. A body of water or marsh land is significant if the area exceeds one-quarter square mile and cannot be traversed by conventional land rescue vehicles. To the extent practicable, the plan shall provide for rescue vehicles with a combined capacity for handling the maximum number of persons that can be carried on board the largest air carrier aircraft that the airport reasonably can be expected to serve.

(g) Each certificate holder shall --

(1) Coordinate its plan with law enforcement agencies, rescue and fire fighting agencies, medical personnel and organizations, the principal tenants at the airport, and all other persons who have responsibilities under the plan;

(2) To the extent practicable, provide for participation by all facilities, agencies, and personnel specified in

paragraph (g)(1) of this section in the development of the plan;

(3) Ensure that all airport personnel having duties and responsibilities under the plan are familiar with their assignments and are properly trained;

(4) At least once every 12 months, review the plan with all of the parties with whom the plan is coordinated as specified in paragraph (g)(1) of this section, to ensure that all parties know their responsibilities and that all of the information in the plan is current; and

(5) Hold a full-scale airport emergency plan exercise at least once every 3 years.

(h) Each airport subject to 49 CFR part 1542, Airport Security, shall ensure that instructions for response to paragraphs (b)(2) and (b)(6) of this section in the airport emergency plan are consistent with its approved security program.

(i) FAA Advisory Circulars in the 150 Series contain standards and procedures for the development of an airport emergency plan which are acceptable to the Administrator.

Sec. 139.329 - Ground vehicles.

Each certificate holder shall --

(a) Limit access to movement areas and safety areas only to those ground vehicles necessary for airport operations;

(b) Establish and implement procedures for the safe and orderly access to, and operation on, the movement area and safety areas by ground vehicles, including provisions identifying the consequences of noncompliance with the procedures by an employee, tenant, or contractor;

(c) When an air traffic control tower is in operation, ensure that each ground vehicle operating on the movement area is controlled by one of the following:

(1) Two-way radio communications between each vehicle and the tower,

(2) An escort vehicle with two-way radio communications with the tower to accompany any vehicle without a radio, or

(3) Measures acceptable to the Administrator for controlling vehicles, such as signs, signals, or guards, when it is not operationally practical to have two-way radio communications with the vehicle or an escort vehicle;

(d) When an air traffic control tower is not in operation, provide adequate procedures to control ground vehicles on the movement area through prearranged signs or signals;

(e) Ensure that each employee, tenant, or contractor who operates a ground vehicle on any portion of the airport that has access to the movement area is familiar with the airport's procedures for the operation of ground vehicles and the consequences of noncompliance; and

(f) On request by the Administrator, make available for inspection any record of accidents or incidents on the movement areas involving air carrier aircraft and/or ground vehicles.

Sec. 139.331 - Obstructions.

Each certificate holder shall ensure that each object in each area within its authority which exceeds any of the heights or penetrates the imaginary surfaces described in part 77 of this chapter is either removed, marked, or lighted. However, removal, marking, and lighting is not required if it is determined to be unnecessary by an FAA aeronautical study.

Sec. 139.333 - Protection of navaids.

Each certificate holder shall --

(a) Prevent the construction of facilities on its airport that, as determined by the Administrator, would derogate the operation of an electronic or visual navaid and air traffic control facilities on the airport;

(b) Protect, or if the owner is other than the certificate holder, assist in protecting, all navaids on its airport against vandalism and theft; and

(c) Prevent, insofar as it is within the airport's authority, interruption of visual and electronic signals of navaids.

Sec. 139.335 - Public protection.

(a) Each certificate holder shall provide --

(1) Safeguards acceptable to the Administrator to prevent inadvertent entry to the movement area by unauthorized persons or vehicles; and

(2) Reasonable protection of persons and property from aircraft blast.

(b) Fencing meeting the requirements of 49 CFR part 1542 in areas subject to that part is acceptable for meeting the requirements of paragraph (a)(1) of this section.

Sec. 139.337 - Wildlife hazard management.

(a) Each certificate holder shall provide for the conduct of an ecological study, acceptable to the Administrator, when any of the following events occurs on or near the airport:

(1) An air carrier aircraft experiences a multiple bird strike or engine ingestion.

(2) An air carrier aircraft experiences a damaging collision with wildlife other than birds.

(3) Wildlife of a size or in numbers capable of causing an event described in paragraph (a) (1) or (2) of this section

is observed to have access to any airport flight pattern or movement area.

(b) The study required in paragraph (a) of this section shall contain at least the following:

(1) Analysis of the event which prompted the study.

(2) Identification of the species, numbers, locations, local movements, and daily and seasonal occurrences of wildlife observed.

(3) Identification and location of features on and near the airport that attract wildlife.

(4) Description of the wildlife hazard to air carrier operations.

(c) The study required by paragraph (a) of this section shall be submitted to the Administrator, who determines whether or not there is a need for a wildlife hazard management plan. In reaching this determination, the Administrator considers --

(1) The ecological study;

(2) The aeronautical activity at the airport;

(3) The views of the certificate holder;

(4) The views of the airport users; and

(5) Any other factors bearing on the matter of which the Administrator is aware.

(d) When the Administrator determines that a wildlife hazard management plan is needed, the certificate holder shall formulate and implement a plan using the ecological study as a basis. The plan shall --

(1) Be submitted to, and approved by, the Administrator prior to implementation; and

(2) Provide measures to alleviate or eliminate wildlife hazards to air carrier operations.

(e) The plan shall include at least the following:

(1) The persons who have authority and responsibility for implementing the plan.

(2) Priorities for needed habitat modification and changes in land use identified in the ecological study, with target dates for completion.

(3) Requirements for and, where applicable, copies of local, state, and federal wildlife control permits.

(4) Identification of resources to be provided by the certificate holder for implementation of the plan.

(5) Procedures to be followed during air carrier operations, including at least --

(i) Assignment of personnel responsibilities for implementing the procedures;

(ii) Conduct of physical inspections of the movement area and other areas critical to wildlife hazard management sufficiently in advance of air carrier operations to allow time for wildlife controls to be effective;

(iii) Wildlife control measures; and

(iv) Communication between the wildlife control personnel and any air traffic control tower in operation at thc airport.

(6) Periodic evaluation and review of the wildlife hazard management plan for --

(i) Effectiveness in dealing with the wildlife hazard; and

(ii) Indications that the existence of the wildlife hazard,

as previously described in the ecological study, should be reevaluated.

(7) A training program to provide airport personnel with the knowledge and skills needed to carry out the wildlife hazard management plan required by paragraph (d) of this section.

(f) Notwithstanding the other requirements of this section, each certificate holder shall take immediate measures to alleviate wildlife hazards whenever they are detected.

(g) FAA Advisory Circulars in the 150 series contain standards and procedures for wildlife hazard management at airports which are acceptable to the Administrator.

Sec. 139.339 - Airport condition reporting.

(a) Each certificate holder shall provide for the collection and dissemination of airport condition information to air carriers.

(b) In complying with paragraph (a) of this section, the certificate holder shall utilize the NOTAM system and, as appropriate, other systems and procedures acceptable to the Administrator.

(c) In complying with paragraph (a) of this section, the certificate holder shall provide information on the following airport conditions which may affect the safe operations of air carriers:

(1) Construction or maintenance activity on movement areas, safety areas, or loading ramps and parking areas.

(2) Surface irregularities on movement areas or loading ramps and parking areas.

(3) Snow, ice, slush, or water on the movement area or loading ramps and parking areas.

(4) Snow piled or drifted on or near movement areas contrary to §139.313.

(5) Objects on the movement area or safety areas contrary to §139.309.

(6) Malfunction of any lighting system required by §139.311.

(7) Unresolved wildlife hazards as identified in accordance with §139.337.

(8) Nonavailability of any rescue and firefighting capability required in §§139.317 and 139.319.

(9) Any other condition as specified in the airport certification manual or airport certification specifications, or which may otherwise adversely affect the safe operations of air carriers.

(d) FAA Advisory Circulars in the 150 series contain standards and procedures for using the NOTAM system for dissemination of airport information which are acceptable to the Administrator.

Sec. 139.341 - Identifying, marking, and reporting construction and other unserviceable areas.

(a) Each certificate holder shall --

(1) Mark and, if appropriate, light in a manner acceptable to the Administrator --

(i) Each construction area and unserviceable area which is on or adjacent to any movement area or any other area of the airport on which air carrier aircraft may be operated;

(ii) Each item of construction equipment and each construction roadway, which may affect the safe movement of aircraft on the airport; and

(iii) Any area adjacent to a navaid that, if traversed, could cause derogation of the signal or the failure of the navaid, and

(2) Provide procedures, such as a review of all appropriate utility plans prior to construction, for avoiding damage to existing utilities, cables, wires, conduits, pipelines, or other underground facilities.

(b) FAA Advisory Circulars in the 150 series contain standards and procedures for identifying and marking construction areas which are acceptable to the Administrator.

Appendix 7:

TCAS Installation and Use

(Taken from FAA Advisory Circular 120-55B)

TCAS systems are now implemented in Part 121 operations and other applications. Approval of TCAS for FAA Type Certification (TC) or Supplemental Type Certification (STC) is comprehensively addressed in AC 20-131, Airworthiness and Operational Approval of Traffic Alert and Collision Avoidance Systems (TCAS II) and Mode S Transponders, as amended. This AC provides information for US air carriers, aircraft and TCAS manufacturers, various inspectors, foreign air carriers operating in U.S. airspace, and other aviation organizations regarding standard means acceptable to the FAA to establish and ensure continued compliance with 14 CFR as related to TCAS. This information is intended to promote timely and comprehensive program implementation, to encourage development of standard practices for the application of TCAS, and to provide for suitable follow-up to TCAS events.

The use of TCAS in Part 121 operations requires both FAA airworthiness certification and operational approval. Airworthiness certification of TCAS refers to an FAA approval of changes in an aircraft's type design by amendment to a TC or issuance of an

STC. Operational approval pertains to changes to training and maintenance programs, manuals, operational procedures, Minimum Equipment Lists (MEL), and other areas necessary for safe and effective TCAS use and the qualification of aircrews through the approved training programs. An airworthiness TC/STC of a TCAS system alone does not constitute operational approval for use of TCAS under provisions of Part 121.

Air carriers must ensure appropriate flight crew TCAS qualification. The flight crew must demonstrate proficiency knowledge of TCAS concepts, systems, and procedures and cognitive, procedural, and motor skills necessary to properly respond to TCAS advisories.

First-time TCAS qualification must be accomplished for each airplane type. Qualification may be accomplished during initial, transition, or upgrade ground and flight training programs with appropriate differences. By this method, TCAS information will be integrated with other curriculum elements and modules. First-time TCAS qualification may also be accomplished as a stand-alone module of ground and flight training. Recurrent TCAS qualification will be accomplished during recurrent ground and flight training. Recurrent ground training will be a stand-alone module. However, TCAS will be fully integrated with the recurrent flight training during proficiency training (PT) or line-oriented flight training (LOFT). For first-time and recurrent TCAS qualification, an instructor will accomplish evaluation of TCAS objectives during training. There are no formal TCAS evaluation requirements for flight testing or checking. However, routine TCAS operations will be included in all evaluation environments and check airmen/examiners should include TCAS as a routine discussion item. Principal Operations Inspectors (POI) may give an operator credit when training is conducted by another operator or training center stand-alone TCAS program for first-time qualification if that program has been approved by the FAA and if aircraft, TCAS system, procedures, and other relevant factors or circumstances are the same or equivalent to those of the operator seeking credit. The POI should consult with the appropriate division of AFS, National Simulator Evaluation Team (NSET), or the assigned AEG about the suitability of a proposed program for a particular version of TCAS or aircraft type.

TCAS knowledge must be evaluated with written, oral, or computer-based instructional tests. Combinations of these methods may be used if the required body of knowledge is completely covered. For any of these methods, a passing grade of 90% must be achieved. First-time qualification in any type airplane must include evaluation of all knowledge areas. For recurrent training, complete coverage of the knowledge requirements must be completed every 36 months. TCAS skills (maneuvers) must be evaluated by an instructor for first-time TCAS qualification in each type airplane. This evaluation may be accomplished by an instructor in a qualified Flight Training Device (FTD), Simulator, or Computer-Based Instructional System (CBI) approved for each maneuver. For recurrent training, all maneuvers must be provided during training in any 36 month period. Recurrent training is desirable in an FTD or Simulator approved for the maneuvers. LOFT programs using simulators equipped with TCAS should be enhanced by an interactive TCAS. In addition, LOFT programs should consider proper crew vigilance for aircraft that may not be transponder or Mode C equipped. Advisories accomplished during LOFT are creditable toward first-time or recurrent qualification.

Individual crewmember TCAS knowledge and skills must be evaluated prior to TCAS use. Acceptable means of initial assessment include the evaluation by an authorized instructor or check airman using written, computer-based, or oral tests, and a simulator, FTD, or CBI system capable of depicting traffic encounters. TCAS recurrent training should be integrated into and/or conducted in conjunction with other established recurrent training programs. Recurrent training for TCAS must include both ground and flight (maneuver) and should address any significant issues identified by line operating experience (OE), system changes, procedural changes, or unique characteristics such as the introduction of new aircraft/display systems or operations in airspace where high numbers of Traffic Advisories (TA) and Resolution Advisories (RA) have been reported. Recurrent TCAS checking should be incorporated as an element of routine proficiency training. When TCAS-equipped aircraft are used during line or route checks, check airmen should routinely incorporate proper TCAS use as a discussion item. LOFT programs using simulators equipped with TCAS should be enhanced by interaction with TCAS. In addition, LOFT programs should consider proper crew vigilance for aircraft which may not be transponder or Mode C equipped. CRM programs should address effective teamwork in responding to TCAS events with emphasis on the following areas: the crew should conduct preflight briefings on how

TCAS advisories will be handled; and the proper reaction to a TA by the pilot flying (PF) and the pilot-not-flying (PNF).

Airplane flight manuals, operating manuals, maintenance manuals, general policy manuals, other manuals, publications, or written material (such as operating bulletins that may relate to TCAS use) must be appropriately amended to describe TCAS equipment, procedures, and operational policies according to the appropriate regulation.

Operators who have aircraft with TCAS differences in displays, controls, procedures, or involved with interchange operations must account for those TCAS differences. This is accomplished as part of an approved differences training program in accordance with Part 121, or as otherwise specified in applicable FAA FSB reports concerning crew qualification pertaining to a particular airplane type. Operators should address any TCAS issues that may be unique to their particular route environment, aircraft, procedures, or TCAS display and control features. Examples include the following: air carriers having takeoffs or landings outside of the reference TCAS performance envelope (for example, airport elevations outside of the range between sea level and 5,300 feet mean sea level (MSL), or temperatures outside the range of International Standard Atmosphere ((ISA) ±50° F) should advise crews of appropriate procedures and precautions regarding RA compliance. To ensure proper response to TCAS in limiting performance conditions (for example, TCAS RA during takeoff climb or in final landing configuration at high altitude airports such as Mexico City and La Paz), specific procedures or training may be needed, unless these situations can be adequately addressed by bulletin or manual information; air carriers should describe the use of TA-only mode of operation when required on certain aircraft with an engine failure; TCAS flight level or absolute display of traffic altitude on a traffic display should not be used during operations when the altimeter is set to zero relative to the intended field of landing field elevation (QFE).

Operationally, those skills addressed and the guidance provided on TCAS training should be followed and implemented by each operator electing to use TCAS II and apply the appropriate 14 CFR. TCAS is intended to serve as a backup to visual collision avoidance, application of right-of-way rules, and air traffic separation service. For TCAS to work as designed, immediate and correct crew response to TCAS advisories is essential. Delayed crew response or reluctance of a flight crew to adjust the aircraft's flight path as advised by TCAS due to Air Traffic Control

(ATC) clearance provisions, fear of later FAA scrutiny, or other factors could significantly decrease or negate the protection afforded by TCAS. Flight crews are expected to respond to TCAS in accordance with the following guidelines when responding to alerts: respond to TAs by attempting to establish visual contact with the intruder aircraft and other aircraft which may be in the vicinity. Coordinate to the degree possible with other crewmembers to assist in searching for traffic. Do not deviate from an assigned clearance based only on TA information.

For any traffic that is acquired visually, continue to maintain or attain safe separation in accordance with current regulations and good operating practices; when an RA occurs, the PF should respond immediately by direct attention to RA displays and maneuver as indicated, unless doing so would jeopardize the safe operation of the flight or the flight crew can assure separation with the help of definitive visual acquisition of the aircraft causing the RA. By not responding to an RA, the flight crew effectively takes responsibility for achieving safe separation. In so choosing, the following cautions should be considered: the traffic may also be equipped with TCAS and it may maneuver in response to an RA that has been coordinated with your own TCAS; the traffic acquired visually may not be the same traffic causing the RA; visual perception of the encounter may be misleading. Unless it is unequivocally clear that the target acquired visually is the one generating the RA and there are no complicating circumstances, the pilot's instinctive reaction should always be to respond to RAs in the direction and to the degree displayed; satisfy RAs by disconnecting the autopilot, if necessary, using prompt, positive control inputs in the direction and with the magnitude TCAS advises. To achieve the required vertical rate (normally 1,500 feet per minute (fpm) climb or descent), first adjust the aircraft's pitch. Then refer to the vertical speed indicator (VSI) and make all necessary pitch adjustments to place the VSI in the green arc.

Operators have the following general responsibilities regarding TCAS: assure follow-up and evaluation of unusual TCAS events; and periodically assess TCAS training, checking, and maintenance programs to ensure their correctness, pertinence, timeliness, and effectiveness.

ATC responsibilities relating to TCAS are: controllers will

not knowingly issue instructions that are contrary to RA guidance when they are aware that a TCAS maneuver is in progress. When an aircraft deviates from its clearance in response to an RA, ATC is still responsible for providing assistance to the deviating aircraft as requested until the pilot informs ATC that the RA conflict is clear; and the aircraft has returned to the previously assigned altitude; or alternate ATC instructions have been issued and acknowledged.

Operators and manufacturers are encouraged to develop procedures to ensure effective identification, tracking, and follow-up of significant TCAS-related events, as appropriate. Such procedures should focus on providing useful information to: properly assess the importance of TCAS events; follow up on information related to specific TCAS events, as necessary; and keep the industry and FAA informed of the performance of TCAS in the NAS and in international operations.

Pilots should make the following reports for TCAS TAs and RAs, as necessary. Upon query from ATC, or after a deviation from an ATC clearance, make radio communications as appropriate to report a response to a TCAS advisory. The FAA will not initiate enforcement action solely on the basis of a TCAS event. Letters of investigation will not be sent to pilots involved in a TCAS-related deviation provided: the aircraft was equipped with TCAS, the system was operable, and the equipment was turned on at the time of the event; the pilots have properly operated their aircraft in compliance with ATC clearances prior to the TCAS-related deviation; and the pilots have successfully completed their air carrier's FAA-approved TCAS training program Certain foreign carriers are required to have TCAS installed when operating in U.S. airspace. Foreign air carriers are not required to install and use TCAS for any aircraft or operations taking place outside of the US 12 nm territorial limit, even though separation services may be provided by a US ATC facility (for example, in oceanic airspace). Various states may abide by ICAO guidance found in ACAS Standards.

An appropriate Mode S transponder must be installed and operated on a suitable code specified by ATC during flight in US airspace. In addition, a valid unique aircraft-specific Mode S address must be assigned to the airplane, and the Mode S transponder must be set to this address. Valid addresses are those consistent with the ICAO Mode S address allocation plan contained in appendix C, part I, ICAO Annex 10, and plan of the State of registry for the specific aircraft. The unique address, when properly set, may not be altered, set to a duplicated address, or set to an address that potentially interferes with ATC or TCAS safety functions (for example, must not be set to all "ones" or all "zeros," or the country address must not be set without the unique aircraft specific address). This guidance is appropriate for non-US registered or US-registered aircraft operated by a foreign air carrier in US airspace.

A TCAS II System capable of coordinating with TCAS units must be installed. The TCAS system must be operated in an appropriate TCAS mode during flight in US airspace. Training and procedures for use of TCAS as specified by ICAO, this AC, or other equivalent criteria acceptable to FAA must be used when operating in U.S. airspace. Unsafe conditions or performance related to TCAS operation which potentially could affect continued safe operations in the US NAS must be reported to the FAA POI within 10 business days of the time that such a hazard is identified. In order to keep everyone informed during a TCAS maneuver, radio communication should be in terminology common to all parties on the frequency regarding a TCAS RA. The following phraseology is suggested and should contain: (1) name of the ATC facility, (2) aircraft identification (ID), and (3) nature of the TCAS deviation. When a flight crew receives a TCAS RA to either climb or descend from their assigned altitude, or the RA otherwise affects their ATC clearance or their pending maneuver or maneuver in progress, the crew should inform ATC when beginning the excursion from clearance or as soon as workload allows in the following manner:

"XYZ Center, (Aircraft ID), TCAS Climb/Descent"

EXAMPLE:

"New York Center, Quantum 321, TCAS Climb"

"Cleveland Center, Universal 602, TCAS Descent"

Following such a communication, the designated air traffic facility is not required to provide

approved standard separation to the TCAS maneuvering aircraft until the TCAS encounter is cleared and standard ATC separation is achieved. If workload permits, traffic information should be provided. When the RA is clear, the flight crew should advise ATC that they are returning to their previously assigned clearance or subsequent amended clearance in the following manner:

"ABC Center, (Aircraft ID), clear of conflict, returning to assigned altitude."

EXAMPLE:

"Boston Center, Northern 429, clear of conflict, returning to assigned altitude."

When the deviating aircraft has renegotiated its clearance with ATC, the designated air traffic facility is expected to resume providing appropriate separation services in accordance with FAA

Ground training must cover the following areas:

a. General Concepts of TCAS Operation. TCAS ground training should cover, in general terms, TCAS theory to the extent appropriate to ensure proper operational use. Aircrews should understand basic concepts of TCAS logic, CPA, tau, altitude separation thresholds for the issuance of RAs, as well as the relationship between displayed traffic information and issuance of TAs and RAs. The ground training program should address the following:

(1) The meaning of TAs and preventive versus corrective RAs;

(2) Increase, reversal, crossing, and weakened RAs;

(3) That TCAS II assures separation from Mode C equipped aircraft;

(4) The detection and protection provided by TCAS against altitude reporting and non-altitude reporting intruders;

(5) That the system detects multiple aircraft;

(6) TCAS-to-TCAS coordination;

(7) The potential impact of not following RAs;

(8) TCAS surveillance range versus display range;

(9) When an intruder will not be displayed;

(10) TCAS on ground performance; and

(11) The continued applicability of the see-and-avoid concept.

b. Expected Flight crew Response and Level of Protection Provided by TCAS. Academic training should explain the normal, expected pilot response to TAs, RAs, use of displayed traffic information to establish visual contact, and constraints on maneuvering based solely on TAs.

c. TCAS General Limitation. There are several system, operational, and/or performance limitations which should be understood that apply to all aircraft types. System limitations include the inability of TCAS to detect non-transponder equipped aircraft, no RAs issued for traffic without an altitude reporting transponder, etc. Operational limitations include some RA inhibit altitudes, certain RAs being inhibited by aircraft performance constraints, the inability to comply with an RA due to aircraft performance limitations after an engine failure, and appropriate response to RAs in limiting performance conditions, such as during heavy weight takeoff or while en route at maximum altitude for a particular weight.

d. ATC Communication and Coordination. Training should discuss communication and coordination with ATC related to or following a TCAS event, when to contact ATC, and accepted TCAS phraseology.

e. TCAS Equipment Components Controls, Displays, Audio Alerts, and Annunciations. Academic training should include a discussion of TCAS terminology, symbology, operation, and optional controls and display features, including any items particular to an air carrier's implementation or unique to its system.

f. Interfaces and Compatibility with Other Aircraft Systems. Training should discuss the role of the Mode S transponder with a correct, discreet address installed, radar altimeter inputs to TCAS, and weather radar/EFIS interfaces, including any items particular to an air carrier's implementation or unique to its system.

g. Aircraft Flight Manual (AFM) Information. AFM provisions should be addressed, including information on TCAS modes of operation; normal and atypical flight crew operating procedures; and response to TAs, RAs, and any AFM limitations.

h. MEL operating provisions.

i. Appropriate pilot response to TCAS RAs and TAs, ATC clearance compliance, nuisance

alerts, and other such issues.

j. The air carrier's TCAS event reporting policies for flight crews.

k. Flight crew procedures for reporting TCAS malfunctions or irregularities, if not otherwise addressed by routine maintenance procedures of that operator.

b. Classroom Training. An understanding of TCAS operation and the criteria used for issuing TAs and RAs may be assessed using the following: Objectives and Criteria. This training should address the following topics:

a. System Operation.

(1) Objective: Demonstrate knowledge of how TCAS functions.

(2) Criteria: The pilot must demonstrate an understanding of the following functions:

(a) Surveillance:

i. TCAS interrogates other transponder-equipped aircraft within a nominal range of 14 nautical miles (nm).

ii. TCAS surveillance range can be reduced in geographic areas with a large number of ground interrogators and/or TCAS II equipped aircraft.

(b) Collision Avoidance:

i. TAs can be issued against any transponder-equipped aircraft which responds to the ICAO Mode C interrogations, even if the aircraft does not have altitude reporting capability.

ii. RAs can be issued only against aircraft that are reporting altitude and in the vertical plane only.

iii. RAs issued against a TCAS-equipped intruder are coordinated to ensure complementary RAs are issued.

b. Advisory Thresholds.

(1) Objective: Demonstrate knowledge of the criteria for issuing TAs and RAs.

(2) Criteria: The pilot must be able to demonstrate an understanding of the methodology used by TCAS to issue TAs and RAs and the general criteria for the issuance of these advisories to include:

(a) TCAS advisories are based on time to closest point of approach (CPA) rather than distance. The time must be short and vertical separation must be small, or projected to be small, before an advisory can be issued. The separation standards provided by Air Traffic Services are different from the miss distances against which TCAS issues an alert.

(b) Thresholds for issuing a TA or RA vary with altitude. The thresholds are larger at higher altitudes. (c) The TA tau threshold (trigger point) varies from 15 to 48 seconds before the projected CPA and the RA tau threshold varies from 15 to 35 seconds.

(d) RAs are chosen to provide the desired vertical miss distance at CPA. As a result, RAs can instruct a climb or descent through the intruder aircraft's altitude.

c. TCAS Limitations.

(1) Objective: To verify the pilot is aware of the limitations of TCAS.

(2) Criteria: The pilot must demonstrate a knowledge and understanding of the TCAS limitations including:

(a) TCAS will neither track nor display non-transponder equipped aircraft, nor aircraft not responding to TCAS Mode C interrogations.

(b) TCAS will automatically fail if the input from the aircraft's barometric altimeter, radio altimeter, or transponder is lost.

NOTE: In some installations, the loss of information from other onboard systems such as an Inertial Reference System (IRS) or Attitude Heading Reference System (AHRS) may result in a TCAS failure. Individual operators should ensure their pilots are aware of what types of failures will result in a TCAS failure.

(c) An intruder aircraft within 380 feet AGL (nominal value) may or may not be displayed by your TCAS (i.e., declared to be airborne or on the ground, respectively) depending upon whether the intruder is Mode S or ATCRBS Mode C equipped and whether your TCAS-equipped aircraft is airborne or on the ground.

(d) TCAS may not display all proximate transponder-equipped aircraft in areas of high density traffic.

(e) Because of design limitations, the bearing displayed by

TCAS is not sufficiently accurate to support the initiation of horizontal maneuvers based solely on the traffic display.

(f) Because of design limitations, TCAS will not track intruders with a vertical speed in excess of 10,000 fpm. In addition, the design implementation may result in some short term errors in the tracked vertical speed of an intruder during periods of high vertical acceleration by the intruder.

(g) Ground Proximity Warning System (GPWS) warnings and wind shear warnings take precedence over TCAS advisories. When either a GPWS or wind shear warning is active, TCAS aural annunciations will be inhibited.

d. TCAS Inhibits.

(1) Objective: To verify the pilot is aware of the conditions under which certain functions of TCAS are inhibited.

(2) Criteria: The pilot must demonstrate a knowledge and understanding of the various

TCAS inhibits including:

(a) Increase Descent RAs are inhibited below 1,450 (±100) feet AGL.

(b) Descend RAs are inhibited below 1,100 (±100) feet AGL.

(c) All RAs are inhibited below 1,000 (±100) feet.

(d) All TCAS aural annunciations are inhibited below 500 (±100) feet AGL. This

includes the aural annunciation for TAs.

(e) Altitude and configuration under which Climb and Increase Climb RAs are inhibited.

Know if your aircraft type issues Climb and Increase Climb RAs when operating at the aircraft's certified ceiling. If your aircraft type provides RA Climb and Increase Climb commands at certified ceiling, the commands are to be followed.

NOTE: In some aircraft types, Climb or Increase Climb RAs are never inhibited.

e. Use of Controls.

(1) Objective: To verify the pilot can properly operate all TCAS and display controls.

(2) Criteria: Demonstrate the proper use of controls including:

(a) Aircraft configuration required to initiate a Self Test.

(b) Steps required to initiate a Self Test.

(c) Recognizing when the Self Test was successful and when it was unsuccessful. When the Self Test is unsuccessful, recognizing the reason for the failure, and if possible, correcting the problem.

(d) Recommended usage of range selection. Low ranges are used in the terminal area and the higher display ranges are used in the en-route environment and in the transition between the terminal and en-route environment.

(e) If available, recommended usage of the Above/Below mode selector. Above mode should be used during climb and the Below mode should be used during descent.

(f) Recognition that the configuration of the display does not affect the TCAS surveillance volume.

(g) Selection of lower ranges when an advisory is issued to increase display resolution.

(h) If available, selection of the display of absolute altitude instead of relative altitude and the limitations of using this display if a barometric correction is not provided to TCAS.

(i) Proper configuration to display the appropriate TCAS

information without eliminating the display of other needed information.

NOTE: The wide variety of display implementations make it difficult to establish more definitive criteria. When the training program is developed, this general criteria should be expanded to cover specific details for an operator's specific display implementation.

f. Display Interpretation.

(1) Objective: To verify a pilot understands the meaning of all information that can be displayed by TCAS.

(2) Criteria: The pilot must demonstrate the ability to properly interpret information displayed by TCAS including:

(a) Other traffic, i.e., traffic within the selected display range that is not proximate traffic, or causing a TA or RA to be issued.

(b) Proximate traffic, i.e., traffic that is within 6 nm and ±1200 feet.

(c) Non-altitude reporting traffic.

(d) No bearing TAs and RAs.

(e) Off-scale TAs and RAs. The selected range should be changed to ensure that all available information on the intruder is displayed.

(f) Traffic advisories. The minimum available display range which allows the traffic to be displayed should be selected to provide the maximum display resolution.

(g) Resolution advisories (Traffic Display). The minimum available display range of the traffic display which allows the traffic to be displayed should be selected to provide the maximum display resolution.

(h) Resolution advisories (RA Display). Pilots should demonstrate knowledge of the meaning of the red and green areas displayed on the RA display and when the green areas will and will not be displayed. Pilots should also demonstrate an understanding of the RA display limitations (i.e., if a vertical speed tape is used and the range of the tape is less than 2,500 fpm, an Increase Rate RA cannot be properly displayed).

(i) If appropriate, awareness that Navigation Displays oriented on Track-Up may require a pilot to make a mental adjustment for drift angle when assessing the bearing of proximate traffic.

NOTE: The wide variety of display implementations will require the tailoring of some criteria. When the training program is developed, these criteria should be expanded to cover details for an operator's specific display implementation.

g. Use of the TA-Only Mode.

(1) Objective: To verify that a pilot understands the appropriate times to select the TA-only mode of operation and the limitations associated with using this mode.

(2) Criteria: The pilot must demonstrate the following:

(a) Knowledge of the operator's guidance for the use of TA-only.

(b) Reasons for using this mode and situations in which its use may be desirable. If TA-only is not selected when an airport is conducting simultaneous operations from parallel runways separated by less than 1,200 feet, and to some intersecting runways, RAs can be expected.

(c) The TA aural annunciation is inhibited below 500 feet AGL. As a result, TAs issued below 500 feet AGL may not be noticed unless the TA display is included in the routine instrument scan.

(d) When this mode is selected, TAs will be issued at the time an RA is normally issued.

h. Crew Coordination.

(1) Objective: To verify the pilot adequately briefs other crew members on how TCAS advisories will be handled.

(2) Criteria: The pilot must demonstrate their preflight briefing addresses the procedures that will be used in responding to TAs and RAs including:

(a) Division of duties between PF and PNF.

(b) Expected call-outs.

(c) Communications with ATC.

(d) Conditions under which an RA may not be followed and who will make this decision.

NOTES:

1. Different operators have different procedures for conducting preflight briefings and for responding to TCAS advisories. These factors should be taken into consideration when implementing the training program.

2. The operator must specify the conditions under which an RA need not be followed, reflecting advice published by the State Civil Aviation Authority. This should not be an item left to the discretion of a crew.

3. This portion of the training may be combined with other training such as CRM.

i. Reporting Requirements.

(1) Objective: To verify the pilot is aware of the requirements for reporting RAs to the controller and other authorities.

(2) Criteria: The pilot must demonstrate the following:

(a) The use of the phraseology contained in PANS-RAC (ICAO DOC. 4444).

(b) Where information can be obtained regarding the need for making written reports to various states when an RA is issued. Various states have different reporting requirements and the material available to the pilot should be tailored to the airline's operating environment.

j. TCAS Flight Training (Maneuver). The scenarios included in the maneuver training should include corrective RAs, initial preventive RAs, maintain rate RAs, altitude crossing RAs, increase rate RAs, RA reversals, weakening RAs, and multi-aircraft encounters. Training must provide pilots the opportunity to reach the TCAS proficiency indicated in the following TA and RA response objectives. This proficiency may be assessed and certified by a TCAS qualified instructor.

k. TA Responses.

(1) Objective: To verify the pilot properly interprets and responds to TAs.

(2) Criteria: The pilot must demonstrate the following:

(a) Proper division of responsibilities between the PF and PNF. The PF should continue to fly the airplane, and be prepared to respond to any RA that might follow. The PNF should provide updates on the traffic location shown on the TCAS display, using this information to help visually acquire the intruder.

(b) Proper interpretation of the displayed information. Both pilots confirm that the aircraft they have visually acquired is that which has caused the TA to be issued. Use should be made of all information shown on the display, note being taken of the bearing and range of the intruder (amber circle), whether it is above or below (data tag), and its vertical speed direction (trend arrow).

(c) Other available information is used to assist in visual acquisition. This includes ATC party-line information, traffic flow in use, etc.

(d) Because of the limitations that may exist with various display systems, the PF should not maneuver the aircraft based solely on the information shown on the TCAS display.

No attempt should be made to adjust the current flight path in anticipation of what an RA would advise.

(e) When visual acquisition is attained, right-of-way rules are used to maintain or attain safe separation. No unnecessary maneuvers are initiated. The limitations of making maneuvers

based solely on visual acquisition, especially at high altitude or without a definite horizon, are understood.

1. RA Responses.

(1) Objective: To verify the pilot properly interprets and responds to RAs.

(2) Criteria: The pilot must demonstrate the following:

(a) Proper division of responsibilities between the PF and PNF. The PF responds to the RA with positive control inputs, when required, while the PNF provides updates on the traffic location and cross-checks between the traffic display and monitors the response to the RA. Proper CRM should be used.

(b) Proper interpretation of the displayed information. The pilot recognizes the intruder causing the RA to be issued (red square on display). Pilot responds appropriately.

(c) For corrective RAs, the response is initiated in the proper direction within 5 seconds of the RA being displayed.

(d) Recognition of the initially displayed RA being modified. Response to the modified RA is properly accomplished.

i. For Increase Rate RAs, the vertical speed is increased within 2 1/2 seconds of the RA being displayed.

ii. For RA reversals, the vertical speed is reversed within 2 1/2 seconds of the RA being displayed.

iii. For RA weakening, the vertical speed is modified to initiate a return towards the original clearance within 2 1/2 seconds of the RA being displayed.

iv. For RAs which strengthen, the vertical speed is modified to comply with the revised RA within 2 1/2 seconds of the RA being displayed.

(e) Recognition of altitude crossing encounters and the proper response to these RAs.

(f) For preventive RAs, the vertical speed needle remains outside the red area on the RA display.

(g) For Maintain Rate RAs, the vertical speed is not reduced. Pilots should recognize that a Maintain Rate RA may result in crossing through the intruder's altitude.

(h) If a decision is made to not follow an RA, no changes in the existing vertical speed are made in a direction opposite to the sense of the displayed RA. Pilots should be aware that if the intruder is also TCAS-equipped, the decision to not follow an RA may result in a decrease in separation at CPA because of the intruder's RA response.

(i) When the RA weakens, pilot initiates a return towards the original clearance, and when Clear of Conflict is annunciated, pilot completes the return to the original clearance.

(j) The controller is informed of the RA as soon as time and workload permit, using the standard phraseology.

(k) When possible, an ATC clearance is complied with while responding to an RA. For example, if the aircraft can level at the assigned altitude while responding to a Reduce Climb or Reduce Descent RA, it should be done.

(l) If pilots simultaneously receive instructions to maneuver from ATC and an RA which are in conflict, the pilot should follow the RA.

(m) Knowledge of the TCAS multi-aircraft logic and its

limitations. For example, TCAS only considers intruders which it believes to be a threat when selecting an RA. As such, it is possible for TCAS to issue an RA against one intruder which results in a maneuver towards another intruder that is not classified as a threat. If the second intruder becomes a threat, the RA will be modified to provide separation from that intruder.

(n) The consequences of both responding to, and not responding to, an RA.

m. Characteristics of Training Equipment Suitable for Maneuver Training.

(1) Acceptable Characteristics. Flight training devices, simulators, and CBIs must have certain characteristics to be effective. This is due to the interactive nature of TCAS, the variety of encounter scenarios possible, the immediate and standardized pilot response required, and the instant and correct display interpretation that is necessary. Thus, training equipment used for TCAS flight training should have the following characteristics:

(a) The ability to functionally represent TCAS displays, controls, indications, and annunciations;

(b) Ability to depict selected traffic encounter scenarios, including TCAS display and audio advisories;

(c) Ability to show proper TCAS reaction to depicted scenarios and advisories; and

(d) Ability to interactively respond to pilot inputs regarding TCAS advisories, including responses to RAs displayed on relevant vertical speed and pitch indicators.

(2) Simulator and TCAS Fidelity. For a particular TCAS, maneuver training may be accomplished in simulators or training devices that represent the specific aircraft or an aircraft that has similar characteristics. For the purposes of TCAS maneuver training, simulators or training devices may use simplified TCAS algorithms or displays and do not require

TCAS logic or a TCAS processor. TCAS displays do not have to be identical, but must be functionally equivalent to the air carrier operator's specific aircraft in use.

(3) Training Device or Simulator Approval. Training devices or simulators meeting FAA criteria are qualified by the NSET and approved for use by the POI. Any one or combination of the following devices or simulators which meet characteristics of paragraph m(1) may be used:

(a) Level A through D simulators;

(b) Level 2 through 7 FTDs; or

(c) Dedicated TCAS training devices acceptable to the FAA, including those devices described in FAA Order 8400.10, Air Transportation Operations Inspector's Handbook, volume 3, paragraph 443, Aircraft Systems Integration Training, which are shown to be suitable for TCAS training and approved by the POI.